The Utterly, Completely, and Totally

Useless Science
Fact-O-Pedia

About the Author

An award-winning and internationally recognized expert medical writer and author of several books, **Wendy Leonard, PhD, MPH,** has been using her gift for translating medical jargon into lay-friendly language for over 25 years. A member of the American Public Health Association, Wendy is also the Medical News Anchor for Montgomery Undercover on MocoVox.com, and resides in the Agricultural Reserve, north of Washington, DC.

About the Illustrators

Illustrator **Matt Ryan** and gag writer **Steve Kanaras** began their comic and cartoon collaborations while attending high school in Granby, Connecticut. Their popular "Junk Food" comic strip regularly appears in newspapers, websites, and other publications. Ryan has been illustrating cartoons and teaching the craft for nearly two decades. He lives in Granby with his wife and two daughters. Steve writes jokes and performs comedy, but has not yet convinced anyone to marry him. He lives in Enfield, Connecticut with the Goddess, Artemis—a five-year-old black Labrador retriever.

The Utterly, Completely, and Totally

Useless Science Fact-O-Pedia

A Startling Collection of Scientific Trivia
You'll Never Need to Know

Wendy Leonard, PhD, MPH

HarperCollins*Publishers*

Copyright © 2012 by HarperCollins*Publishers*

This 2012 edition published for Barnes and Noble, Inc.
by HarperCollins*Publishers*

3 5 7 9 10 8 6 4 2

Interior illustrations © Steve Kanaras and Matt Ryan

Wendy Leonard asserts the moral right to
be identified as the author of this work

A CIP catalog record for this book is
available from the British Library

ISBN: 978-0-00-792779-1

Printed and bound in China
by South China Printing Co. Ltd.

Acknowledgments

I've never been given an acknowledgments page by a publisher before, so this is beyond thrilling for me! Why? You know when you're watching an awards show, and the winner goes on and on about how "none of this would have been possible without so-and-so and so-and-so, etc.?" I get that now.

Labor of love that it is notwithstanding, writing a book is an enormous and insanely time-consuming undertaking that requires not only unconditional understanding from all those people that one simply cannot focus on to the level they deserve, it also requires loving emotional support (from same said people!) whenever one's energy, focus, and basic brain processing skills begin to fall into a dark abyss (read: whenever it totally tanks). And that's just part of it!

To that end, to follow are the amazing folks for whom I must give a shout out and my deepest, most heartfelt thanks for a multitude of reasons, including—but not limited to—giving me some incredible topic ideas; candidly assessing the "cool factor" of various topic points; providing me with unwavering love and encouragement whenever I was feeling crispy; and frankly (whether knowingly or not), for helping keep me sane … well, sane-ish!

Thank you: Linda Ramsdell, John, Christa and Jeremy Mobley, Elise Stigliano, Steve Kanaras, Kait Rowe, Raimi Kellner, Noam Laden, Kimberly Mazzocchi, Lill Becker, Dave O'Brien, Mark Schaffer, Tim Welch, Tim Elliott, Michelle Bowen, Gabe Sullivan, Nina Chazen, Nika, Mariel and Ariana Leonard, Donna Lewis, Craig MacEachern, Leo Eaton, Michael Grant, Wade Grubic, Stewart Waller, Denise Frank-Conneen, Muriel Leonard, Elaine Silver, my sibs Julie, Steve, and Deb (and their cool kids), my mom (Renee), who taught me about the Writing Faeries, my dad (Jerry), whose mantra (well, one of his many mantras) is: "Interesting, but upon what is that based?"; the Cousin John Band, whose "Jellyfish" CD kept the Writing Faeries dancing on my keyboard, and Jeannine Dillon, the Publishing Director of HarperCollins*Publishers*, who not only really "gets" me and my nerdy sensibilities, but also made sure my personal interjections (usually in parentheses) were kept in my book: Jeannine, you rock!

I'd also like to give specific thanks to Dr. Chris Portier for providing me with late-breaking CDC news on lead levels; Chris Dionigi and Diane Leonard for their invasive species expertise; Tim Howe, for his expertise on Brasswinds; Christa Mobley, for all things equestrian; and most of all, to my incredible husband, Dr. Chris Leonard, who created and updated my life-saving Excel spreadsheet, made sure I ate, and was my number-one resource for clarifying all things scientific (like, "Honey, you ever heard of a voltaic pile?"). As one of my fabulous nieces is famous for saying, whenever she wants to convey someone is super intelligent, "Oh, he's Uncle Chris smart!" Thank you so much, honey. Mwah!

Yes, I see the hook coming: And to all those who I forget to thank, please know that you are deeply appreciated, too! XOXO to the Moon and back!

Contents

Introduction

Of course we don't know what we're doing, that's why we call it research!
—Albert Einstein

"Wendy, please try to mitigate your general tendency towards getting easily distracted by shiny objects." This was publicly posted by the amazing and brilliant Dr. Regina A. Galer-Unti, PhD, CHES, the chairperson of my PhD dissertation thesis. I smile from ear to ear every time I think about that!

Now, she's not the first person to share this sentiment with me, and hopefully, she won't be the last! As it's my insatiable, albeit wandering, curiosity that compels me to think: "Really? Is there any scientific literature to back up that claim?" Thus, I drill deeper and deeper, and invariably along the way more points of contention (and general coolness) pop up, which is how this book came to be!

Here's the thing: My preferred path is not the one of least resistance, it's the one with the most potential for interesting detours!

Of note, and to the dismay of pretty much every teacher and professor I've ever had, I don't "do" outlines. My goal isn't to fill in the blanks of preconceived notions and ideas. My goal is to discover (and then share with others) uber-awesome, interesting scientific and medical information and their associated totally cool factoids in such a way that people are veritably compelled to repeat the aforementioned factoids at small social gatherings! Besides, outlines get in the way of hearing the Writing Faeries.

Now, to supplant any possible lingering musings as to why it is that this book discusses everything from amino acids to geomagnetism to zombie ants, and then, in the middle of debunking a spurious theory about plant-animal mutualism and the extinct dodo bird (*Raphus cucullatus*) section, I interject:

✒ The dodo was a character in Lewis Carroll's *Alice in Wonderland*. In the 1865 book, the dodo is believed to be a self-caricature of Carroll, whose real name was Charles Lutwidge Dodgson. Purportedly, because of a stutter, he was known to introduce himself as "Do-do-dodgson."

That's how my brain works!

And just for the record, after having a series of neurological tests a few years back, a world-renowned doctor offered me this conclusion: "Wendy, you have abnormal brain waves," to which I respectfully responded, "Compared to what? I've never been tested before; it's not like you have a baseline." Then, without so much as blinking, the rather avuncular doctor replied, "Compared to the rest of the human species."

And whenever I share this story, no one ever seems surprised.

I truly hope, with all my heart, that you enjoy reading this book (and the incredible cartoons provided by Steve Kanaras and Matt Ryan) as much as I did writing it!

Kindly, and with ubuntu,
Wendy

Amino Acids

- Amino acids are the building blocks of proteins, which are the machinery of life on our planet.

- Amino acids are so named because they contain an amine group and a carboxylic acid group connected by a central "alpha" carbon.

- A peptide bond—which is the primary linkage of all protein structures—is formed when the amine group of one amino acid is connected to the carboxylic acid of another amino acid.

- Every amino acid has a mirror-image molecule called an isomer, which has an L-form and a D-form.

- The term "L-form" comes from the Latin word *levo*, meaning left; "D-form" comes from the Latin word *dextro*, meaning right.

- Nearly all naturally occurring amino acids are L-form (left-handed) amino acids.

- Chemically synthesized amino acids (made in a lab) are usually a 50/50 mixture of L-form (left-handed) and D-form (right-handed) isomers.

- Contrary to popular belief, taurine is not an amino acid. It contains a sulfonic acid group, not a carboxylic acid group. Of note, the human body can manufacture taurine from other amino acids. However, cats cannot—which is why taurine is an essential ingredient in cat food.

A

Antioxidants

୶ Antioxidants such as Vitamin A, C, and E, Lycopene, Lutien, and Glutathione are substances that protect cells from unstable molecules called "free radicals." These antioxidants are abundant in fruits and vegetables, as well as in other foods including nuts, grains, and some meats, poultry, and fish.

୶ Free radicals can damage your DNA (deoxyribonucleic acid). This is a problem because your genes—which are pieces of DNA—are responsible for providing your cells with the necessary instructions to function, including when to grow and divide, and when to die.

A ୶ Damaged DNA has been linked to some forms of cancer because cancer cells don't know when to stop reproducing and die.

୶ In humans, the most common form of free radicals is oxygen.

୶ When an oxygen molecule (O_2) becomes electrically charged, or "radicalized," it tries to steal electrons from other molecules, causing damage to DNA and other molecules. Over time, this damage may become irreversible and lead to disease.

୶ According to the National Cancer Institute, antioxidants essentially "mop up" free radicals, meaning they neutralize the electrical charge and prevent the free radical from stealing electrons from other molecules.

"A" Phobias

- **Alektorophobia:** Fear of chickens
- **Alliumphobia:** Fear of garlic
- **Allodoxaphobia:** Fear of opinions
- **Aibohphobia:** Fear of palindromes (my personal favorite)
- **Amathophobia:** Fear of dust
- **Amaxophobia:** Fear of riding in a car
- **Ablutophobia:** Fear of washing or bathing
- **Acousticophobia:** Fear of noise
- **Acrophobia, Altophobia:** Fear of heights
- **Aerophobia:** Fear of drafts, air swallowing, or airborne noxious substances
- **Aeronausiphobia:** Fear of vomiting secondary to airsickness
- **Agateophobia:** Fear of insanity
- **Agliophobia:** Fear of pain
- **Agoraphobia:** Fear of the outdoors, crowds, or uncontrolled social conditions
- **Agrizoophobia:** Fear of wild animals
- **Agyrophobia:** Fear of streets or crossing the street
- **Aichmophobia:** Fear of needles or pointed objects
- **Ailurophobia:** Fear of cats

A

Apollo 11
(Moon Landing)

🐦 On July 20, 1969, Apollo 11 astronauts Neil Armstrong—the crew Commander—and Edwin "Buzz" Aldrin, Jr.—the Lunar Module Pilot—were the first people to land on the Moon. Neil Armstrong was the first person to set foot on the Moon.

🐦 Michael Collins—the Command Module Pilot—was part of the crew, but stayed in orbit around the Moon in the command module, while Neil and Buzz descended to the Moon's surface in the Lunar Excursion Module (LEM).

🐦 Neil Armstrong actually flubbed his famous scripted line. He was supposed to say: "That's one small step for a man, one giant leap for mankind," but he left out the "a," and instead he said: "That's one small step for man, one giant leap for mankind." So, instead of stating that a simple, everyday action of one man had monumental implications for all humanity, he instead made the nearly contradictory statement that suggested that a small step by the human race resulted in a monumental achievement by humankind! But, no do-overs in space!

🐦 NASA "covered this up" by stating that static from the broadcast of the statement made the "a" inaudible. Based upon the actual recorded transmission from that day, this doesn't appear to be the case.

🐦 Years later, Neil Armstrong purportedly said: "Damn, I really did it. I blew the first words on the Moon, didn't I?"

~ That said, I think it's safe to say that syntax error or no syntax error, Neil Armstrong's famous Moon landing statement resonates as an amazing, beautiful, life-changing, brilliant moment in the history of not just the United States, but for all humanity.

Apples

- Apples are members of the rose family.

- The science of apple growing is called "pomology."

- There are over 7,500 varieties of apples; 2,500 varieties of apples are grown in the United States.

- Be patient: Apple trees take four to five years to produce their first fruit.

- Contrary to popular belief, 80% of an apple's soluble fiber comes from the pectin—the white, fleshy part of the apple—not the skin!

- Apples have more quercetin than any other fruit. Quercetin is a heart-healthy flavonoid that possesses outstanding antioxidant and anti-inflammatory properties.

- The crab apple tree is the only apple tree native to North America.

- Apples ripen six to ten times faster at room temperature than if they were refrigerated.

- The term Adam's Apple (which refers to the larynx or voice box that becomes particularly visible in boys once puberty hits) is a reference to the Biblical story in Genesis, where Adam purportedly got a piece of the forbidden fruit stuck in his throat.

- In colonial times in the United States, apples were called "winter bananas."

Asian
Longhorned Beetles

🐦 The Asian longhorned beetle (ALB), also known as *Anoplophora glabripennis*, is a cerambycid beetle—a round-headed, wood-boring species native to Asia.

🐦 In China, the name for the beetle translates to the "starry sky" or "starry night" because of its glossy black body covered with irregular white spots, and its white-and-black-banded antennae.

🐦 The ALB is dangerous to trees. All four stages of the beetle's life cycle damage the host trees. Most insect borers are considered "secondary pests" because they attack only after a plant has been weakened or killed by another stress. ALB, however, is a "primary pest" able to attack and develop in fairly healthy trees and kill them.

A

🐦 The ALB prefers hardwood tree species like maple species—including Norway, sugar, silver, red, and box elder maple—and horsechestnut, birch, buckeye, elm, willow, and poplars.

🐦 Most damage is done during the larval stages, when eggs are injected under the bark surface where they hatch into larvae. Larvae tunnel under the bark and destroy the tree's vascular system, which disrupts the sap flow of infested trees. Older larvae tunnel into the heartwood where their feeding slowly destroys the structural integrity of trees. ALB-infected trees are slowly killed over a three to five year period, although it may take longer.

🐝 The beetle completes most of its life cycle inside the host tree, with adults emerging in late spring. Adult beetles feed on twigs, leaf petioles, and primary leaf veins.

🐝 Because the beetle spends most of its life cycle feeding on the dead heartwood of the tree—where chemical treatments can't penetrate—the only known effective control method for ALB eradication is by the identification and removal of the infested trees.

Aspirin
(Acetylsalicylic Acid)

❧ Aspirin derives from the Latin word *salix*—which means willow tree. As far back as 400 BC, the Greek physician Hippocrates was advising patients to chew on the bark of the White Willow to reduce fever and inflammation.

❧ In 1899, Aspirin was the number-one-selling drug worldwide! Back then, Aspirin was a brand name, coined by Bayer (of course!), so the "A" was capitalized.

❧ In the Middle Ages, the demand for wicker furniture was so great that Europeans stopped using willow bark remedies. And get this: In some places, using willow bark for medicinal purposes was actually forbidden!

❧ It wasn't until 1838 that the Italian chemist Raffaele Piria successfully converted salicin—the name of the natural chemical found in the bark of the White Willow—into salicylic acid in his laboratory. Piria also converted salicin into a sugar.

❧ Fifteen years later, in 1853, French chemist Charles Frederic Gerhardt synthesized salicylic acid into acetylsalicylic acid, albeit in a form that was way too impure and unstable for real-life application.

A

❧ There has been controversy over which scientist was responsible for developing aspirin into a pure and stable form. Bayer maintains that a chemist named Dr. Felix Hoffmann holds that honor.

❧ Prior to the 1970s, we didn't know aspirin's mechanism of action. We had no idea *how* it worked—only that it did work! So how *does* aspirin work? In a nutshell, it works by inhibiting the production of prostaglandins.

Asteroids

❧ Asteroids are irregular or spheroid in shape and are essentially comprised of chunks of rock. They are thought to be remnants and debris from the original giant cloud of gas and dust that condensed—and ultimately created—the Sun, the planets, and moons in our solar system some 4+ billion years ago.

❧ There are over 100,000 asteroids in the main belt between Mars and Jupiter. This is not the same belt as the Kuiper Belt, which is located in the outermost fringes of the solar system.

❧ Thousands of the biggest (known) asteroids in the Kuiper Belt have been given their own names, such as Hermes—which, at its closest approach, came within 482,805 miles (777,000 km) of the Earth. The largest known asteroid is named Ceres. It's approximately 590 miles (950 km) wide. In 2004, an asteroid named Toutatis came within 961 miles (1.5 million km) of Earth. That's about four times the distance to the Moon.

A

❧ Asteroids dramatically range in size from just a few feet to several miles in diameter. Small asteroids are called meteoroids. [*See* Meteoroids, Meteors, and Meteorites, *page 144.*]

❧ Toutatis, which was discovered in 1989, is shaped like a potato and is 2.9 miles (4.6 km) long. It's named after the Celtic god of war and growth.

❧ Many scientists believe that it was the impact of an asteroid hitting the Earth 65 million years ago that caused, or at least contributed to, the extinction of the dinosaurs. [*See* K-T Extinction, *page 122.*]

Axolotls
(Pronounced: *ACK-suh-LAH-tuhl*)

A type of amphibious salamander, the axolotl's physical genome—its genetic blueprint—is the largest of any known species on the planet, including humans. In fact, it's ten times the size of the human genome!

Injured axolotls have the amazing ability to regenerate fully functioning organs and body parts like their arms, legs, eyes, heart, spine, and even portions of their brain. How is this possible? Axolotls can dedifferentiate their own cells (yes, we're talking stem cells) just like they did as embryos, still inside the egg.

Scientists have been studying and publishing papers about axolotl biology since the 1800s. Understanding the mechanisms behind their regeneration capabilities would be the scientific breakthrough of the century (any century!). It would be truly life altering for millions of people.

The axolotl was once an abundant species thriving in the ancient Aztec waters, but today the axolotl's only remaining natural habitat is Lake Xochimilco (pronounced *SO-chee-MILL-koh*). The lake has been drained, and it's now a series of meandering canals, partly comprised of untreated sewage water. (Yuck ... that can't be good for them.)

Axolotls live up to 15 years (in captivity); average an impressive 12 inches (30.5 cm) long; come in a beautiful array of colors including black, pink, gold, olive, brown, gray, white (albino), and multicolored; and, are the only amphibian that can reach sexual maturity when still a larva.

Bacteria

🦠 Bacteria are single-cell organisms. They've been around for billions of years and are the oldest living organisms on Earth.

🦠 While the human body is made up of about one trillion cells, you have about ten trillion bacterial cells in or on your body at any given moment.

🦠 Bacteria are instrumental in keeping us alive, digesting food, making vitamins, and educating the immune system to keep "bad" microbes out—these are "good" bacteria.

🦠 "Bad" bacteria make humans sick, such as cholera, Lyme's disease, and E. coli.

🦠 Contrary to popular belief, while bacteria are primitive, "simple" organisms, they actually talk to each other via chemical languages!

🦠 In fact, similar bacteria—referred to as intra-species bacteria—communicate with each other in their own language so they know who and how many are in the neighborhood, which dictates what they may or may not do. This is called "quorum sensing."

🦠 Different types of bacteria—referred to as "inter-species bacteria"—have a universal language (a five-carbon molecule) that allows *all* bacteria to communicate with each other! This helps them avoid duplicating jobs! (Very cool, and kind of scary.)

B

Big Bang Theory

 The Big Bang occurred approximately 13.75 (± 0.11) billion years ago.

 It's believed that it took only an instant (a trillion-trillionth of a second!) after the "big bang" for the universe to go from an infinitely dense point to its ever-expanding astronomical grandeur.

 The first galaxies were created about 500 million years after the Big Bang.

 Our solar system was created about eight billion years after the Big Bang.

B The basic ideas of the Big Bang Theory were first published by Belgian priest and astrophysicist, Monsignor Georges Henri Joseph Édouard Lemaître, although these ideas are often inaccurately credited to Edwin Hubble.

 The term "Big Bang Theory" was originally coined by Sir Fred Hoyle as a contrast to his own theory of a steady-state universe.

 The original cause of the Big Bang still eludes scientists.

Black Holes

- Albert Einstein's Theory of General Relativity predicted that:

  - When a massive star dies, it leaves behind a small, dense remnant core.
  - According to his equation, if the core's mass is more than about three times the mass of the Sun, the force of gravity overwhelms all other forces, producing a black hole. He was correct!

- Most black holes form from the remnants of a large star that dies in a supernova explosion.

- As the surface of the star nears an imaginary surface called the "event horizon," Time on the star slows relative to the time kept by observers far away.

- While a star is in the process of collapsing (under the overwhelming influence of gravity), two very cool things happen, from the perspective of an observer from a distance:

  - The passage of Time on the surface of the star will appear to be moving in slow motion as it approaches the "event horizon".
  - Then, once the star's surface reaches the "event horizon," Time will appear to totally stand still (no movement, nothing; just frozen).
  - However, from the perspective of the star, it just keeps collapsing!
  - It's all relative!

B

❧ A Princeton physicist named John Wheeler didn't coin the term "black hole" until 1967.

❧ Just for the record—and as predicted by Einstein—smaller stars don't become black holes. Instead, they become dense neutron stars, which aren't massive enough to trap light.

Blood

🌿 Blood is formed in your bone marrow, which is the soft, spongy center of your bones.

🌿 Blood is actually one of your connective tissues; it's a liquid tissue and the only liquid tissue in your body.

🌿 Blood makes up around 7% of the weight of the human body.

🌿 Blood cells circulate in the human body for about 120 days.

🌿 Blood helps your body regulate body temperature. It does this by redistributing heat to skin to facilitate cooling via the evaporation process.

🌿 While blood in the arteries is bright red, blood in the veins is more of a dark maroon shade because it carries impurities back to the kidneys and liver for disposal.

🌿 When looking at our skin, the blood looks bluish because of light refraction and other factors but blood really is a reddish color. And just for the record, our veins are actually white!

B

Brain

🐦 The average adult human brain weighs about 3 pounds (1.4 kg). For comparison, a whale brain weighs 8 to 15 pounds (3.6 to 6.8 kg); an elephant brain weighs 8 to 11 pounds (3.6 to 5 kg); a chimp brain weighs about 0.75 pounds (0.35 kg); and a dolphin brain weighs about the same as a human brain, about 3.5 pounds (1.6 kg).

🐦 Although it makes up just 2% of the body's weight, the brain uses around 20% of its energy.

🐦 Albert Einstein's brain weighed only 2 pounds (0.9 kg)—less than the average adult brain. Einstein's brain had significantly more neurons—a.k.a. brain cells—packed into those 2 pounds!

B

🐦 The right side of the human brain thinks in pictures and is "present" focused, meaning it's focused on what's going on right here and right now. The left side of the brain—which thinks in language—processes linearly and methodically and is focused on the past and the future.

🐦 Exclusive to *Homo sapiens*, one of the many responsibilities of the prefrontal cortex of the brain—located just behind the forehead—is to function like an experience simulator. Yes, this amazing region of the brain is what allows us to imagine things without having to actually *do* the thing. For example, we don't have to taste the combination of fish, onion, butterscotch, and ice cream to know it will be disgusting; we don't have to drive off a cliff to know that doing so is very bad idea.

Brasswinds
(Far from "Bore-ing")

🐦 Ever wonder why a trumpet and a trombone have very clear, straight tones, while French horns have a rounder, mellower sound? It's because of a construction concept called "bore."

🐦 Much like a rifle bore, a brass instrument's bore describes the inner diameter of a tube.

🐦 Trumpets and trombones are "cylindrical bore," meaning that aside from the mouthpiece and the flare of the bell, their bore stays the same diameter for the length of the instrument.

🐦 Horns, on the other hand, have "conical" bores, meaning the tube increases in diameter throughout its length, resulting in the warmer, rounder sound.

🐦 Thus, a trombone can have that very cool slide, which would be physically impossible for a conical bore instrument! For a slide to work, the tube has to be a constant diameter.

B

Buckyballs

❧ Buckyballs are spherical molecules of pure carbon with the chemical formula C60 (the "C" standing for carbon). They resemble a tournament soccer ball with 20 hexagon and 12 pentagon faces.

❧ The proper name for C60 is Buckminsterfullerene, so named because it looks like the geodesic dome structure made famous by the American architect R. Buckminster Fuller.

❧ Buckyballs are the third naturally occurring form of pure carbon after graphite and diamond.

❧ Buckminsterfullerene was named the 1991 Molecule of the Year by *Science* magazine.

B

❧ Specially modified buckyballs have wide applications in the fields of electronics (superconductivity), medicine (drug delivery and medical imaging), chemistry (catalyst), and renewable energy (hydrogen storage in fuel cells).

Butterflies *and* Moths

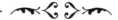

🦋 Butterflies and moths are both of the order *Lepidoptera*. Butterflies and moths are *holometabolous*, meaning that they undergo a complete metamorphosis from egg to caterpillar and from chrysalis to adult.

🦋 There are far less species of butterflies than moths. Butterflies and skippers (a skipper is a type of butterfly that has hooked-shaped antennae) make up 6–11% of *Lepidoptera* order, while moths make up 89–94%.

🦋 The largest and rarest butterfly in the world is the Queen Alexandra Birdwing (*Ornithoptera alexandrae*). From the rain forests of Papua New Guinea, it has a wingspan of 11 inches (28 cm).

🦋 The smallest known butterflies are the blues (*Lycaenidae*), which are found in North America and Africa. They have wingspans from ¼ to ½ inches (6 to 13 mm). The largest known moths are the Atlas moths (*Saturniidae*) with wingspans as large as 12 inches (30.5 cm).

🦋 The smallest known moths are from the pygmy moth family (*Nepticulidae*), with wingspans as small as 3/32 of an inch.

🦋 It's a myth that if you touch the wing of a butterfly or moth and powder rubs off that the butterfly or moth will no longer able to fly. The powder is actually tiny scales, which are modified hairs that naturally shed throughout the insect's lifetime.

🦋 The word *Lepidoptera* comes from the Greek word *lepis* meaning "scale," and *pteron*, meaning "wing."

B

Butterflies *versus* Moths

* **Wings**
  * Butterflies tend to fold their wings vertically up over their backs.
  * Moths usually hold their wings in a tent shape to hide their abdomen.
  * Butterflies usually have more colorful patterns on their wings.
  * Moths typically have drab-colored, smaller wings.

* **Anatomy**
  * Moths have a wing-coupling device called a "frenulum," which joins the forewing to the hind wing, so the wings can work in unison during flight.
  * Butterflies don't have frenulums.

B

* **Behavior**
  * Butterflies are primarily diurnal, meaning they fly in the daytime. But some butterflies are crepuscular, meaning they fly at dawn and dusk.
  * Moths are generally nocturnal, meaning they fly at night. There are some moths that are diurnal, like the buck moth.

* **Cocoon/Chrysalis**

Note: Both cocoons and chrysalides are protective coverings for the pupa, which is the intermediate stage between the larva and adult.

  * Moths make a cocoon, which is wrapped in a silk covering.
  * Butterflies make a chrysalis, which is hard, smooth, and has no silk covering.

Carbon

🌰 Carbon is an element that occurs in organic compounds, like living things, and in many inorganic compounds, like limestone, coal, and petroleum.

🌰 Carbon is the fourth most common element in the universe (after hydrogen, helium, and oxygen).

🌰 Carbon is the 15th most common element in Earth's crust.

🌰 Just behind oxygen, carbon is the second most common element in the human body.

🌰 The word "carbon" comes from the Latin word *carbo*, meaning "coal".

C

🌰 Carbon atoms can attach themselves to one another to form long chains and rings. No other atoms of other elements can attach themselves to one another like carbon atoms. Carbon's "favorite" atom to attach to is hydrogen, but there are many others that it attaches to, including oxygen, nitrogen, fluorine, chlorine, iodine, and sulfur, to name a few.

🌰 There are more than ten million known carbon compounds, each with their own distinct chemical and physical properties and characteristics.

🌰 Plastics are made from carbon polymers, which are long-chain molecules.

❧ Carbon atoms can form chains that are literally thousands of atoms long. It can also form rings, rings with chains, some with branches, and even cross links!

❧ Carbon compounds such as chlorofluorocarbons are associated with ozone depletion, as well as with the greenhouse effect.

❧ All organisms absorb carbon from their environment until they die. Carbon dating—also referred to as radiocarbon dating—utilizes a special method of measuring the naturally occurring isotope Carbon-14—which is produced by cosmic rays in the stratosphere and upper troposphere—to glean a reasonable estimate of its age. [*See* Radiocarbon Dating, *page 196.*]

❧ The diamond—one of the hardest substances on Earth—and graphite—one of the softest substances on Earth—are both 100% pure carbon.

Chocolate

෴ Raw and processed chocolate come from the beans of the *Theobroma* cacao tree. We call them cocoa "beans," but they aren't beans at all—they're seeds from the tree. So technically speaking, chocolate is a fruit!

෴ *Theobroma* cacao is Latin for "food of the gods."

෴ The average cocoa pod—which weighs about one pound (0.45 kg)—contains anywhere between 20 to 60 seeds.

෴ There's documented use of cocoa as far back as around 1100 BC. However, we know that cocoa has been cultivated for at least three millennia in places such as Mexico, and Central and South America.

C

෴ Scientific research has shown that dark chocolate—which consists of a 60% cocoa minimum—is beneficial for the heart. Moderate dark chocolate consumption has also been shown to increase insulin.

෴ The antioxidants found in dark chocolate—called "flavonoids"—have been shown to increase "good" HDL cholesterol, lower "bad" LDL, and lower blood pressure.

෴ White chocolate is antioxidant/flavonoid-empty, as it contains no cocoa solids.

෴ Milk chocolate is also antioxidant/flavonoid empty because the cocoa solids content of commercially produced types ranges from 20–34%, depending upon the brand—not the required minimum of 60%.

Cholesterol

꩜ A type of fat, cholesterol comes from just two sources: your body and food.

꩜ Your liver and other cells in your body make about 75% of the body's total blood cholesterol; the other 25% comes from the foods you eat.

꩜ Food sources of cholesterol are only found in animal products, like butter, eggs, beef, pork, lamb, duck liver, and whole-milk dairy products.

꩜ The word "cholesterol" comes from the Greek word *chole*—meaning "bile, gall"—and the Greek word *stereos*—meaning "solid, stiff."

C ꩜ Gallstones develop when bile contains too much cholesterol and not enough bile salts.

꩜ For optimal health, the body needs a small amount of cholesterol to perform key functions, including producing hormones, vitamin D, and bile—an acid that helps your body digest fat.

꩜ HDL is the "good" kind of cholesterol and stands for high-density lipoprotein. HDL is referred to as the good cholesterol because it helps remove artery-clogging cholesterol from the bloodstream.

👟 LDL, the "bad" kind of cholesterol, stands for low-density lipoprotein. If your bloodstream has too much of the cholesterol that's packaged in your LDL, the surplus is dumped into your arteries, which is why LDL is referred to as the bar cholesterol. Over time, this may result in atherosclerosis—hardening of the arteries—which is the most common cause of heart disease.

👟 An HDL of 60 mg/dL (milligrams per deciliter of blood) and higher provides some protection against heart disease. An LDL of less than 100 mg/dL is considered optimal.

👟 According to the latest research, having a total cholesterol level of less than 200 mg/dL is considered desirable, as statistically, this puts you at lower risk for coronary heart disease.

👟 Having a total cholesterol level of 240 mg/dL is considered high. A person with this level has more than twice the risk of coronary heart disease as someone whose cholesterol is below 200 mg/dL.

Colloidal Silver
(A Bad Case of the Blues)

❧ Silver—the lustrous metal we know and love—is not toxic. However, there is a form of silver called "colloidal silver" that consists of tiny silver particles suspended in liquid. If ingested, this is bad. Make that *really* bad.

❧ Despite a variety of medical claims on the Internet that colloidal silver is a "cure-all," colloidal silver is neither safe nor effective for treating any disease or condition!

❧ Silver, no matter what the form, is NOT a nutritionally essential mineral. Thus, claims that you may suffer from a silver "deficiency" are entirely unfounded.

C

❧ In fact, ingesting silver or even being exposed to high levels for a long period of time may result in a condition called arygria, which causes a blue-gray discoloration of the skin, body tissues, and organs. This condition is typically not treatable nor is it reversible. In other words, your skin will be bluish ... forever! (Although, being blue is not technically harmful.)

❧ As described by the CDC's Agency for Toxic Substances and Disease Registry, "Argyria is a permanent effect, but it appears to be a cosmetic problem that may not be otherwise harmful to health."

❧ The FDA and the Federal Trade Commission have taken action against a number of companies—including some companies that sell products over the Internet—for making false claims about the benefits of colloidal silver products.

❧ There are no oral silver-containing FDA-approved medications—neither prescription nor over-the-counter products. There are, however, safe and effective FDA-approved silver-containing topical preparations for external use, such as burn cream.

Comets

  ✒ Comets are essentially balls comprised of rock and ice. The number of comets in the solar system is believed to be in the trillions.

  ✒ Unlike asteroids, comets have "tails," some of which can be millions of miles long.

  ✒ In the course of its elliptical orbit, a comet's tail grows as it approaches the Sun.

  ✒ When a comet heats up, and thereby evaporates because of its proximity to the Sun, it produces the comet's telltale jets of gas and dust to form the tail. The Sun illuminates this tail trail, which is what causes it to glow.

  ✒ Some comets make repeated visits, while others do not.

  ✒ Some of the more "famous" comets include Halley's, Shoemaker-Levy, and Hale-Bopp.

  ✒ Hale-Bopp was one of the brightest comets ever seen from Earth.

C

Computers

❧ In 1823, Charles Babbage began constructing the mechanical "Difference Engine," which is generally recognized as the first multipurpose or programmable computing device.

❧ For the 1890 census, the U.S. Census Bureau employed a mechanical computer using punch-card equipment for input that was invented by Herman Hollerith.

❧ In 1911, Hollerith's company merged with another to form a third company, which became International Business Machines (IBM) in 1924.

❧ In 1943, work began on the Electronic Numerical Integrator and Computer (ENIAC), which was the first general-purpose programmable electronic computer and was used to solve ballistics problems during World War II.

C

❧ The ENIAC contained over 18,000 vacuum tubes.

❧ ENIAC was displayed to the public on February 14, 1946, at the Moore School of Electrical Engineering at the University of Pennsylvania.

❧ In 1964, the IBM 360 became the first computer to be mass-produced.

Corals

~❧ While people tend to think of coral as hard, stonelike inanimate rock formations that sometimes gets turned into pretty jewelry, corals are actually living, breathing marine animals. Their class is Anthozoa, their phylum is Cnidaria).

~❧ Almost all corals are colonial organisms, meaning they're composed of thousands of tiny, individual invertebrate animals called polyps—which are tubular saclike animals.

~❧ Polyps are also related to sea anemones and jellyfish. They're amazing little creatures that secrete the mineral calcium carbonate, which, over time, accumulates and becomes the amazing reefs we know and love.

C

~❧ One of the largest corals is the *Fungia*—the mushroom coral—which is a solitary coral that can grow to be 10 to 12 inches (25 to 30 cm) in diameter. Colonial corals are much smaller; their polyps range between 0.04 and 0.12 inches (1 to 3 mm) in diameter.

~❧ Similar to our sense of smell and taste, coral polyps can detect certain substances like sugars and amino acids, which enable corals to detect their prey, such as plankton and small fish.

~❧ Most corals get the majority of their nutrition via their symbiotic relationship with single-celled algae called zooxanthellae—which live within the coral polyp's tissues. [*See* Zooxanthellae, *page 304.*]

❧ Polyps are generally nocturnal feeders, and the stomach cavities of colonial corals are interconnected, meaning food obtained by one polyp can be passed to other polyps in the colony!

❧ While corals don't have "brains," they do have a simple nervous system called a nerve net that extends from their mouth to their tentacles.

❧ Corals can reproduce both sexually and asexually. Some corals are hermaphroditic, meaning they have both male and female reproductive cells, and others are either male or female. Both sexes can occur in a colony, or a colony may consist of individuals of the same sex. Progressive little creatures!

Coriolis Effect

🐦 Earth's rotation causes objects moving over its surface—like missiles or hurricanes—to veer to the right in the Northern Hemisphere and to the left in the Southern because Earth is constantly moving beneath these objects in the direction of its daily rotation. This is the Coriolis effect.

🐦 The Coriolis effect can be observed in any rotating frame of reference such as the spinning Earth or a merry-go-round.

🐦 It is a myth that the Coriolis effect causes bathtubs or sinks to drain one direction in the Northern Hemisphere and the other in the Southern. The power of the effect is far too small to overcome the other factors involved in such events, such as residual currents or the vessel's shape.

🐦 The mathematical expression for the magnitude of the Coriolis effect appeared in an 1835 paper by French scientist Gaspard-Gustave Coriolis, as part of the explanation of the theory of water wheels.

🐦 The Coriolis effect influences left or right movement, but it also influences up and down. This is known as the Eötvös effect.

🐦 With the Eötvös effect, eastward-traveling objects are deflected upward because they feel lighter, and westward-traveling objects are deflected downward because they feel heavier.

🐦 The Coriolis effect has a strong influence on air circulation in a developing tropical cyclone. These rarely form along the equator because the Coriolis effect is very weak there.

DNA
(Deoxyribonucleic Acid)

🖎 DNA is our genetic blueprint, the fundamental building block for an individual's entire genetic make-up.

🖎 Nearly every cell in your body has the same DNA. Most DNA is located in the cell nucleus—where it is called nuclear DNA—but a small amount of DNA can also be found in the mitochondria, where it is called mitochondrial DNA or mtDNA.

🖎 Since human red blood cells don't have a nucleus, they don't contain DNA.

🖎 In 1953, James Watson and Francis Crick became famous for their accurate description of DNA's double-helix model.

🖎 Not so famous is Johann Friedrich Miescher, who, shortly after graduating from medical school, actually *discovered* DNA—which he called "nuclein"—while researching white blood cells in the pus of used hospital surgical bandages. The year was 1869.

🖎 What people tend not to know about DNA is that DNA actually has three naturally occurring structures: A-DNA, B-DNA, and Z-DNA. The double helix of A-DNA and B-DNA twists to the right, and the Z-DNA twists left!

D

Dodo Birds

🐦 The dodo (*Raphus cucullatus*) is a now-extinct, flightless bird related to pigeons and doves. The last confirmed sighting occurred in 1662, on an islet off Mauritius.

🐦 The origin of the name is unclear. One theory suggests that in 1598, Portuguese sailors discovered the previously unknown species, which displayed no fear of humans. They called it *douda*—simpleton in Portuguese.

🐦 The dodo was a character in Lewis Carroll's *Alice in Wonderland*. In the 1865 book, the dodo is said to be a self-caricature of Carroll, whose real name was Charles Lutwidge Dodgson. Purportedly, because of a stutter, he was known to introduce himself as "Do-do-dodgson."

D

🐦 It's a commonly held belief that dodos weighed over 50 pounds/23 kg (thanks to a 17th-century report) but fossil evidence suggests that dodos ranged in size from 24 to 46 pounds (10.6 to 21.1 kg)—the size of a large wild turkey.

🐦 Research suggests that classic illustrations of fat dodos were either exaggerations or were based on overfed specimens. Pictures of fat dodos may also have been based on the dodo's tendency to puff out its feathers.

🐦 In the 1977 publication *Plant-Animal Mutualism: Coevolution with Dodo Leads to Near Extinction of Plant*, the author wrongly suggests that since there were almost no Calvaria trees younger than 300 years (around the same time period as the dodo became extinct), the seeds must have passed through the digestive tract of the dodo to germinate. This is a good example of how correlation and causation are not the same thing.

Dogwood Trees

�</> The Dogwood (*Cornus florida*) is an unusually hard wood and a beautiful spring-flowering tree that bears pink or white flowers.

🌿 The Dogwood is Virginia's state flower *and* its state tree. It is also North Carolina's and Missouri's state flower. Yes, it's a tree and not a flower. But it does "flower" … Hmmmm.

🌿 The deeply ridged and broken bark resembles alligator hide.

🌿 Nothing about the Dogwood tree looks like a dog.

🌿 In the fall, bright red berries called drupes appear at the point where the leaves meet the branches.

🌿 The red berries are the fruit of choice in the fall and winter for the gray squirrel, fox squirrel, bobwhite, cedar waxwing, cardinal, flicker, mockingbird, robin, wild turkey, and woodpecker. The leaves and twigs are choice food for the white-tailed deer.

🌿 Contrary to popular belief, Dogwood berries are not toxic. According to the Poison Control Center of the Children's Hospital of Philadelphia, "Dogwood berries are not toxic when eaten, but there have been reports of rashes after skin contact with the tree."

D

Dolphins

 Dolphins are marine mammals, not fish, and there are 36 different kinds of ocean dolphins and five species of river dolphins.

 Dolphins are descendants of terrestrial mammals, and it is believed that they first successfully ventured into the water roughly 50 million years ago.

 In all 41 species from the *Delphinidae* family, the male dolphins are called bulls, females are called cows, the babies are called calves, and they live in social groups called pods.

 Dolphins have the ability to produce ultrasonic sounds up to 200 Kilohertz (KHz). The highest frequency usually audible by humans is 20 KHz!

 If dolphins lose consciousness, they do NOT breathe and can suffocate as opposed to drowning.

 Dolphins only sleep with one half of their brain at a time. The other hemisphere stays a bit alert, as a protective mechanism. [♪Insert musical theme to *Jaws* here! ♪]

 The killer whale—often referred to as the orca (*Orcinus orca*)—can be mistaken for a whale but is actually the largest dolphin family member.

 Other dolphins wrongly classified as whales: the melon-headed whale (*Peponocephala electra*), pygmy killer whale (*Feresa attenuata*), false killer whale (*Pseudorca crassidens*), long-finned pilot whale (*Globicephala melas*), and short-finned pilot whale (*Globicephala macrorhynchus*).

D

Dust Devils

🐦 Dust devils are well-formed, vertically oriented rotating columns of air that rarely exceed 45 mph (70 km/h).

🐦 Dust devils are formed when a warm air mass near the ground rises, which is why they are common in deserts.

🐦 Unlike tornadoes—which are spawned within thunderstorms when a cloud mass overhead has started rotating and begins to pull warm air in from below—there doesn't need to be a cloud in the sky for dust devils to form.

🐦 Called *chiindii*—meaning ghosts or spirits—by the Navajo Indians, dust devils that spin clockwise are said to be good spirits; dust devils that spin counterclockwise are bad spirits.

D

🐦 Dust devils can range in size from rather small to very large. Small dust devils measure less than 3 feet (1 m) in diameter and dissipate in less than a minute. Large dust devils can measure 300 feet (90 m) with winds reaching 60 mph (100 km/h) and last upward of 20 minutes.

🐦 Dust devils have been seen on Mars, both from the Martian Landers and from orbit. One was estimated to be 22 miles (35 km) high.

Dwarf Planets

❧ In 2006, the International Astronomical Union (IAU) created a classification for dwarf planets. It also created a classification for comets and asteroids, which are now designated as "small solar system bodies."

❧ Dwarf planets are not just distinguished from planets because of their smaller size, they must orbit the Sun, but not other planets, and they must have a robust-enough gravity to be spheroid in shape. This eliminates most asteroids from the category.

❧ When dwarf planets were classified as such, Pluto—formerly known as our solar system's ninth planet—was demoted to dwarf planet status.

D

✥ In addition to Pluto, there are now four additional dwarf planets: Ceres, Eris, Makemake, and Haumea.

✥ Ceres—which used to be classified as an asteroid—is located in the main asteroid belt between Mars and Jupiter.

✥ Eris, Makemake, and Haumea—which orbit our Sun in the Kuiper Belt (sometimes called the Edgeworth–Kuiper Belt)—are located in the outermost fringes of our solar system.

✥ Astronomers believe that Neptune's moons, Triton and Phoebe, originally came from the Kuiper Belt.

✥ Pluto has three moons; Haumea has two moons; and Eris has one moon. And, to the best of our knowledge, none of the dwarf planets have rings.

✥ In terms of atmosphere, Pluto has methane. Eris and Makemake might have methane.

Earth

~~~❦~~~

❧ Formed about 4.5 billion years ago, Earth is the largest of all of the terrestrial planets—the others being Mercury, Venus, and Mars. Earth is 7,926 miles (12,753 km) in diameter and is about 91 million miles (146 million km) from the Sun.

❧ Earth is usually depicted as round in shape, but it is actually an oblate spheroid—meaning it's roundish, but bulges in the center, at the equator. This is due to the centrifugal force of Earth's rotation.

❧ One hundred tons of space dust falls to Earth every day.

❧ With a radius of 763 miles (1,228 km), the innermost core of Earth is about the size of our moon and is comprised mostly of a solid iron-nickel alloy. The outer core, with a radius of 1,408 miles (2,226 km), is liquid. The outer core is crucial to the processes that produce Earth's magnetic field.

❧ For a sense of proportion, Earth's total radius is 3,959 miles (6,371 km).

❧ Earth's mass is known to be approximately 13,173,000,000,000,000,000,000,000, which is about 13 septillion pounds! That's approximately 6,000,000,000,000,000,000,000,000 (6E+24) kg. Or, more specifically: 5.98×10^{24} kg. [Source: NASA].

E

🌿 Earth's atmosphere, which is rather thin, is a mixture of 78% nitrogen, 21% oxygen, 0.9% argon, 0.03% carbon dioxide, and trace amounts of other gases. That said, it's thick enough to insulate us from extreme temperature fluctuations and to block much of the Sun's damaging—and potentially lethal—ultraviolet radiations (better known as UV rays).

🌿 Approximately 70% of Earth's surface is ocean, and the remaining 30% of Earth's surface is land, which is why Earth is often called the "Blue Planet."

🌿 Earth's only known natural satellite is the Moon.

🌿 What's in a name? In Greek mythology, Earth—whose Greek name was *Gaea*—was the mother of the mountains, valleys, streams, and all other terrestrial formations and she was the wife of Uranus. The word "earth" comes from the Old English word *eorthe*, which means "ground." Remember, to ancient peoples, Earth wasn't a planet, she was the very ground we walked upon.

Earthquakes

⟿ The largest recorded earthquake in the world to date had a magnitude 9.5 (Mw) and occurred in Chile on May 22, 1960.

⟿ The largest recorded earthquake in the United States to date had a magnitude 9.2 and struck Prince William Sound, Alaska, on March 28, 1964, which was Good Friday. Not such a good Friday.

⟿ Before electronics, scientists built large spring-pendulum seismometers (like, 15 tons large!) in an attempt to record the long-period motion produced by such quakes. There's one three stories high in Mexico City that's still in operation!

E ⟿ There is no such thing as "earthquake weather." Statistically, there's an equal distribution of earthquakes in all types of weather, including cold weather, hot weather, and rainy weather.

⟿ The San Andreas Fault is NOT a single continuous fault, but rather is actually a fault zone made up of many segments.

⟿ Aristotle (circa 350 BC) wrote about earthquakes! He recognized that "soft ground shakes more than hard rock."

⟿ In 1760, a British engineer named John Michell—one of the first fathers of seismology—accurately wrote in his memoirs that "shifting masses of rock miles below the surface" and the waves of energy they make cause earthquakes.

❧ Each year, the southern California area has about 10,000 earthquakes. Most of them are so small they're not felt.

❧ About 90% of the world's earthquakes occur in the Pacific Ocean's Ring of Fire. [*See* Tsunamis, *page 231.*]

❧ From 1975 to 1995 there were only four states in the U.S. that didn't have any earthquakes! They were Florida, Iowa, North Dakota, and Wisconsin.

❧ The hypocenter of an earthquake is the location beneath Earth's surface where the rupture of the fault begins.

❧ The epicenter of an earthquake is the location directly above the hypocenter on the surface of Earth.

Echinoderms

- Echinoderms are a phylum of marine animals that include sea stars (starfish), brittle stars, sand dollars, sea urchins, sea cucumbers, and crinoids.

- When a sea cucumber is attacked, it may expel some of its internal organs! This may satisfy the predator or scare it off. The sea cucumber then grows another set of organs. (How awesome and really disgusting!)

- Sea stars have the amazing ability to regenerate limbs that have been severed or damaged.

- Echinodermata are the largest phylum without any freshwater or terrestrial forms. There are upward of 7,000 living species of Echinodermata and approximately 13,000 species of echinoderms that are known to be extinct, based upon fossil records.

- The word "echinoderm" derives from the Greek word *echino*—meaning "spiny"—and *derm*, meaning "skin."

- All echinoderms have radial symmetry. This means their appendages point outward from the center, like the spokes on a bike's wheels.

- For achieving locomotion, food and waste transportation, and respiration, echinoderms have hydraulic-pressured water vascular systems!

- The common sea urchin, or European edible sea urchin (*Echinus esculentus*), has been designated "near threatened" on the IUCN Red List of Threatened Species.

Electricity

~<●●>~

∾ While electric current is measured in amperes (amps), electrical potential energy is measured in volts.

🐦 AC/DC: The electrons of AC power—which stands for "alternating current"—have electrons that move back and forth. The electrons of DC power—which stands for "direct current"—move in a single direction.

🐦 Nikola Tesla helped invent AC power, and Thomas Edison helped invent DC power. AC power is safer and can be used over longer distances. The electricity use in homes is AC; batteries use DC.

🐦 The world's biggest source of energy comes from coal. In the United States, coal accounts for about 42% of the four trillion kilowatt hours of electricity generated. After coal, the world's biggest sources of energy come from natural gas (25%), nuclear power (19%), hydropower (8%), wind power (3%), and biomass (1%). Propane, geothermal, and solar power each provide less than 1% of the world's energy.

🐦 In the United States, most of the energy used in homes is for space heating (41%), followed by electronics, lighting, and other appliances (26%), water heating (20%), air-conditioning (8%), and refrigeration (5%).

🐦 Due to the longer heating seasons, the Northeast and Midwest regions of the United States consume the most energy per household, at 123 and 110 million Btu (British thermal units) per household, respectively. The world per capita consumption of energy averages about 81 million Btu.

Energy

❧ Named after James Prescott Joule, the Standard International (SI) unit of measure for energy and work is called the "joule" (J).

❧ There are two types of energy:

    ❧ Stored (potential) energy
    ❧ Working (kinetic) energy

❧ Energy comes in many forms, including heat (thermal), light (radiant), motion (kinetic), electrical, chemical, nuclear, dark, and gravitational.

❧ The United States Energy Information Administration (EIA) estimates that of the annual world energy consumption of 500+ quadrillion Btus, only about 10% is gleaned from renewable energy sources, which include wind, solar, geothermal, biomass, bio-fuels, and hydropower.

❧ Total annual energy consumption in the United States averages about 101 quadrillion Btus, which is about 20% of the world total.

❧ The Conservation Law of Energy states that energy may neither be created nor destroyed. In other words, the sum of all the energies doesn't change; it's a constant.

E

Exercise

꩜⟨ᢒ⟩ᠵᠠ

꩜ Getting 30 minutes or more of daily exercise can:

ᕰ Improve your overall mood.
ᕰ Improve your body's ability to use insulin.
ᕰ Lower blood pressure.
ᕰ Lower "bad" LDL and raise your "good" HDL. [*See* Cholesterol, *page 26.*]
ᕰ Reduce stress.
ᕰ Decrease fatigue.
ᕰ Strengthen your heart.
ᕰ Improve muscle tone and strength.
ᕰ Decrease body fat.
ᕰ Increase energy levels.
ᕰ Reduce your risk for heart disease, high cholesterol, diabetes, high blood pressure, and colon cancer.
ᕰ Help control weight.
ᕰ Help build and maintain healthy bones, muscles, and joints.
ᕰ Help relieve symptoms of anxiety and depression.
ᕰ Increase life expectancy.
ᕰ Increase overall quality of life.

E

Floods

◦❧ The five worst floods in history all occurred in China. These include the 1975 flood in the Hong and Ru rivers; the 1642 flood in the Yellow River; the 1938 flood in the Yellow River; the 1887 flood in the Yellow River; and the 1931 flood in the Yellow River—the worst flood in history. This flood left more than 80 million people homeless, flooded 42,000 square miles (108,880 km²) of land, and had a death toll of between 850,000 to four million.

◦❧ It is has been hypothesized that the Great Biblical flood known as Noah's Ark occurred when the Mediterranean broke through the Bosporus Straits and flooded the Black Sea, circa 5600 BC.

◦❧ In the United States, more than half of all fatalities during floods are auto-related, usually the result of drivers misjudging the depth of water on a road and the force of moving water.

F

◦❧ A car can float in just a few inches of water; just 6 inches (15 cm) of water will reach the bottom of most passenger cars and cause loss of control and possible stalling. Two feet (0.6 m) of rushing water can carry away most vehicles, including sport utility vehicles (SUVs) and pickups.

◦❧ Flash floods are the number-one weather-related killer in the U.S.

◦❧ The principal causes of floods in the Eastern United States and the Gulf Coast are hurricanes and storms; the principal causes of floods in the Western United States are snowmelt and rainstorms.

Flowers
(Largest and Smallest)

- ◆ **Largest bloom in world:**
  - ◇ Perhaps best known for its bloom smelling like rotting meat, the *Rafflesia arnoldii* is the largest bloom in the world.
  - ◇ Found in the rain forests of Indonesia, *Rafflesia arnoldii* can grow to be 3 feet (0.91 m) across and weigh up to 15 pounds (6.8 kg)!
  - ◇ A parasitic plant with no visible leaves, roots, or stem, the *Rafflesia arnoldii* attaches itself to a host plant to obtain the water and nutrients it needs to survive.
  - ◇ The *Rafflesia's* bloom, which only lasts around five to seven days, emits a repulsive scent to attract insects for the purpose of pollination.

F

- ◆ **Smallest bloom in the world:**
  - ◇ Imagine one itty-bitty candy sprinkle! That's the size of an entire watermeal plant!
  - ◇ A member of the *Lemnaceae* family, the watermeal plant (*Wolffia globosa*) averages 1/42 of an inch long and 1/85 of an inch wide.
  - ◇ Okay, I really want you to understand how small I'm talking. Ready? The entire plant weighs about 1/190,000 of an ounce—the same as two grains of table salt! If you tried to fill a thimble with them, you'd need about 5000 plants! If you were to find this rootless plant that happily floats on freshwater lakes and in marshes, you'd think someone had spilled cornmeal!
  - ◇ The flower of the *Wolffia* has single pistil and stamen and thus, not surprisingly, it also produces the world's smallest fruit, called a *utricle*.

Four Forces of Flight

꙳ The four forces of flight are weight, lift, thrust, and drag. These forces make an object move up and down, fast and slow. The amount of each force compared to its opposing force determines how an object moves through the air.

꙳ Weight is the downward force that an aircraft must overcome to fly. Weight is the amount of gravity multiplied by the mass of an object.

꙳ Lift is the opposite force of weight. For an aircraft to move upward, it has to have more lift than weight. Thus, lift is the push that lets something move up.

꙳ Drag is a force that pulls back on something trying to move. The shape of an object also affects drag. Generally speaking, round surfaces usually have less drag than flat ones, and narrow surfaces usually have less drag than wide ones. The existence of drag explains why airplane wings are curved on top and flatter on the bottom. That shape makes air flow over the top faster than under the bottom. As a result, less air pressure is on top of the wing. This lower pressure makes the wing—and the airplane—able to move upward.

F

꙳ Thrust is the opposite force of drag—it's the push that moves something forward. For an aircraft to keep moving forward, it must have more thrust than drag. A small airplane might get its thrust from a propeller; a larger airplane might get its thrust from jet engines; a glider doesn't have thrust and only flies until the drag causes it to slow down and land.

Freezer Burn

~&~ When perfectly good food becomes discolored, parched, frost-covered, and/or has leathery dry spots following a long period in the freezer, we say the food is "freezer-burned." What happened to cause this?

~&~ The freezing process transforms the water molecules within foods to form ice crystals, which is fine because that's how the preservation method works. Here's the thing: Water molecules are drawn to the most hospitable environment, which is the coldest place they can find—the sides of your freezer! It's that migration of water molecules that causes the food to become dehydrated, or freezer-burned.

~&~ Oxygen molecules can also seep into frozen foods, particularly when the food isn't wrapped tightly enough, which dull the color and modifies, or even ruins, the food's flavor.

~&~ Freezer burn is far more likely to occur if the temperature in your freezer fluctuates above 0°F (-17.7°C). This typically occurs when people keep opening and closing the freezer. The standard temperature setting for freezers is usually -10 to 0°F (-23 to -17.7°C).

~&~ And just for the record, there is a limit as to how long foods can be safely frozen. Sooner or later the water molecules will find their way out of the frozen food to a colder place in your freezer.

F

And a few quick tips:

- The faster a food freezes, the better, as this minimizes those undesirable six-sided, snowflake-looking, large ice crystals from forming.
- During thawing, those big ice crystals that form in foods that are frozen slowly damage food cells and dissolve emulsions. This can cause meat to "drip" and lose juiciness, and products such as cream or mayonnaise to separate and appear curdled.
- Ideally, a food that is 2 inches (5 cm) thick should freeze completely in about two hours.
- Never stack packages to be frozen. Instead, spread them out in one layer on various shelves, stacking them only after they're frozen solid.

Frilled-Necked Lizards

🐦 Native to Australia and New Guinea, frilled-necked lizards (*Chlamydosaurus kingii*) are extraordinary creatures.

🐦 Frilled-necked lizards have the coolest way of running. They start on all four limbs and then as they accelerate, they switch to running on their hind legs!

🐦 Also known as frilled lizards and frilled dragons, the frilled-neck lizard's stunning ruffle encircles its neck in a series of perfect pleats, which when startled, angry, frightened, or in courtship, opens in glorious fashion! Their frilled ruffle is also believed to aid in body temperature regulation.

🐦 For optimal camouflage, the color and vividness of the frilled-neck lizard's skin depends upon where they're living! Yellow with black and white markings means they're from Queensland; orange with red, black, and white speckles means they're from the Northern Territory.

🐦 The frilled-necked lizard has a specialized, progressive, five-step defense plan:

> 🐦 Identifying possible danger: Relying on its natural body colors for camouflage, it will slowly cringe down, which makes it look a lot like a stick.
> 🐦 Confirming the threat: He bluffs by opening his mouth wide, which engages the ruffle to open like an umbrella in a blaze of "threatening" color!

F

- When the predator seems unfazed, hissing and jumping toward the predator ensues.
- When this still has no effect, the lizard will up the ante, commencing with repeated lashing of its tail upon on the ground.
- If the aforementioned overtures fail, the frill-necked lizard will opt to run away—preferably up a tree—but will bite predators with his rather large canines, if that's his last option.

Frogs

~&~ Contrary to popular belief, toads aren't reptiles: they're amphibians, just like frogs.

~&~ The icky-sticky secretion on the skin of the Australian green-eyed frog and the growling grass frog is able to successfully neutralize and kill antibiotic-resistant bacteria, including staph infections such as MRSA.

~&~ Frogs don't drink water; they soak it into their body through their skin.

~&~ Frogs have the remarkable ability to see what's straight ahead, to the side, and above them all at the same time.

F ~&~ To help swallow their food, frogs use their eyes! The blinking motion "pushes" their eyeballs downward, thereby creating a bulge in the roof of their mouths, which helps propel the food that's already in their mouths to the back of their throats.

Fungi, Pathogenic

🍃 From the Greek words *pathos*—meaning "suffering, passion"—and *genēs*—meaning "producer of"—a pathogen is defined as any infectious disease-producing agent.

🍃 The study of pathogenic fungi is called medical mycology.

🍃 Phytopathology, which is the study of plant diseases, was an offshoot of medical mycology!

🍃 There are over 100,000 recognized species of parasitic fungi—100 of which are known to be infectious to humans.

🍃 Parasitic fungi are classified into four categories: superficial, subcutaneous, systemic, and opportunistic. Each category reflects a different degree of tissue involvement and mode of entry into the host to create infection:

F

- 🍃 Superficial parasitic fungi cause localized damage to the skin, the hair, and the nails (e.g. Ringworm).
- 🍃 Subcutaneous parasitic fungi cause infections confined to the dermis, subcutaneous tissue, or adjacent structures (e.g. Sporotrichosis).
- 🍃 Systemic parasitic fungi cause infections deep within internal organs (e.g. Histoplasmosis).
- 🍃 Opportunistic parasitic fungi usually cause infections only in the immuno-compromised (e.g. Cryptococcosis). [*See* Pathogens, *page 179, to learn about viral, bacterial, parasitic, and prion pathogens.*]

Fungi, Poisonous
(Mushrooms)

🐦 Contrary to popular belief, there are no (as in *zero*) hard and fast, reliable rules for distinguishing between safe, edible fungi, and those that are poisonous, toxic, and deadly.

🐦 It is NOT true that toxic fungi change color during the cooking process.

🐦 It is NOT true that peeling a toxic fungus makes it safe to eat. Peeled or unpeeled, cooked or uncooked, a poisonous fungus is still poisonous and can often be deadly.

🐦 The deadliest of all fungi is the *Amanita phalloides*, otherwise known as the death cap. Seemingly harmless in appearance, it's about 5 inches (13 cm) across, has a soft, greenish-olive colored cap, and lovely white gills and flesh. They're found in the woodlands, often near oak and beech trees.

🐦 Bearing a disturbingly similar resemblance to the delicious and very edible chanterelle mushroom, the *Cortinarius speciosissimus* is a deadly fungus with a lovely radishlike fragrance. Ranging in color from reddish to tawny brown, its flattish cap ranges in size from ¾ to 3¼ inches (2 to 8 cm), and its gills are rust colored. They're most likely to be seen in the fall … but not always.

꙳ Not to be confused with other edible yellowish fungi such as the chanterelle, the *Paxilus involutus* is deadly. Extremely common in woodland areas—particularly in close proximity to birch trees—this incredibly deadly fungus is about 5 inches (13 cm) wide, has a yellow-brown cap with a rolled rim, and yellow-brownish gills.

꙳ Easily mistaken for the common button mushroom found in grocery stores, the *Amanita virosa*—otherwise known as the "destroying angel"—can be found on lawns, grassy meadows, near trees and shrubs, and in and along the edges of woodlands. The cap of the destroying angel is white, although it can be tan, yellow, or pinkish in the center. At full maturity the cap can be 5 inches (13 cm) wide.

꙳ There is no known antidote for the destroying angel or the death cap. Ninety percent of people who ingest these mushrooms die from liver failure.

Garlic

🌿 A member of the Lily family, garlic—which is also known as the "stinking rose"—has long been known for its many health benefits—a reputation it absolutely deserves!

🌿 Circa 3000 BC, Charak—the father of Ayurvedic medicine—declared that eating garlic maintained the heart and the fluidity of the blood.

🌿 The *Egyptian Codex Ebers* is the oldest preserved medical document in existence, dating back to 1552 BC. Also called the *Ebers Papyrus*, the document claimed that garlic was a treatment for heart disorders. (That's a wow!)

G 🌿 Garlic was considered so sacred to the ancient Egyptians that it was placed in the tombs of Pharaohs.

🌿 Garlic was believed to be so powerful that ancient Greeks and Romans ritually consumed garlic before going to war.

🌿 Hippocrates (c. 460 BC to c. 370 BC) and Pliny the Elder (23 AD to 79 AD) strongly encouraged the consumption of garlic because of its profound health virtues.

🌿 When the Israelites were wandering the desert, they pined for it (Numbers: 11:5).

⤙ Garlic is antioxidant-rich in the form of *allicin* (plus the antioxidant vitamins A and C). A variety of studies clearly suggest that consuming garlic on a regular basis can lower "bad" LDL cholesterol, raise "good" HDL cholesterol, and can prevent, reduce, and even reverse the development of atherosclerosis (hardening of the arteries), which can lead to heart disease, heart attack, and stroke.

⤙ Studies also find that regular consumption of garlic can help lower blood pressure, lower blood sugar levels, reduce the formation of blood clots, relax blood vessels, and improve blood flow.

⤙ To enjoy the antioxidant benefits of allicin, you must first chop, crush, or otherwise "damage" the garlic clove so that a natural chemical reaction occurs that converts the undamaged allicin into heart-healthy, atherosclerosis-fighting allicin.

⤙ To date, no reliable scientific studies confirm the efficacy of garlic as a vampire repellant.

Gemstones

~ Most gemstones contain several elements. For example, sapphires are an aluminum oxide (α-Al2O3), and emeralds are beryllium aluminum silicate with chromium ($Be_3Al_2(SiO_3)_6$). The diamond, however, is all carbon!

~ The sapphire is a corundum, and it comes in every color of the rainbow including purple, blue, and green. However, red "sapphires" are called rubies.

~ While most people think of garnets as red, they also come in every color of the rainbow ... except blue! One of the most valuable is the tsavorite garnet, which is a vivid, radiant green. And like all garnets, the tsavorite has a particularly high refractive index (1.734/44).

~ Pearls, amber, and coral are the only three official gemstones that come from living things.

~ Alexandrite from the Ural Mountains of Russia is a particularly amazing gemstone: It changes colors depending upon the source of light! It's a stunning shade of green, like an emerald, by daylight and then completely changes to the color red, like a ruby, by incandescent light.

~ Truth be told, the terms "precious" and "semiprecious" actually have more to do with an archaic import-export tariff definition rather than their inherent value. For example, tsavorite garnets, tanzanite, and imperial jadeite can easily sell for several thousands of dollars per carat, which is as much or more than most "precious" gemstones, like rubies, emeralds, sapphires, and diamonds.

G

Geomagnetism

🐦 The study of geomagnetism is one of the oldest of the geophysical sciences.

🐦 The earliest ideas about the nature of magnetism are attributed to the pre-Socratic Greek philosopher and the "Father of Science" Thales of Miletus (c. 624 BC to c. 546 BC), although the knowledge of the existence of magnetism dates back to prehistoric times.

🐦 Earth's magnetic field is actually a composite of several magnetic fields generated by a variety of sources; the most important ones being Earth's conducting, fluid outer core, of which more than 90% of the geomagnetic field is generated; magnetized rocks in Earth's crust; fields generated outside Earth by electric currents flowing in the ionosphere and magnetosphere; electric currents flowing in Earth's crust, which is usually induced by varying external magnetic fields, and ocean current effects.

G

🐦 Earth's magnetic field is slowly changing; is different in different places; changes with both location and time; can completely reverse; and changes the way it changes!

🐦 The last time the magnetic field reversed was about 750,000 to 780,000 years ago. If the magnetic field were to reverse today, your compass would point south instead of north.

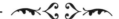

❧ During the past 100 million years, the reversal rates have ranged drastically. Some have occurred within a 5,000-year timespan, and some have occurred within upward of 50 million years.

❧ It's possible that a reversal of the geomagnetic field may affect the migratory behavior of some animals, as many use the geomagnetic field to orient themselves.

❧ And just for the record, the north and south geographic poles and our north and south magnetic poles aren't located in the same place!

G-Force

🐦 The "G" in G-force stands for gravity.

🐦 The term "G-force" is technically a misnomer as it's a measure of acceleration, not force.

🐦 G-force is measured in Gs: One g is equal to the force of acceleration due to gravity near Earth's surface (9.8 m/s^2, or 32 ft/s^2).

🐦 The G-force acting upon an object in any weightless environment is zero Gs. An example of this is a free-fall in a vacuum.

🐦 Pretty much unknown until the advent of airplanes, G-forces were the mysterious reason World War I pilots were blacking out during dogfights.

G

🐦 Today, fighter pilots wear anti-G suits, which have inflatable pants that push the blood that would otherwise pool in their legs back up into their upper body, where it can be circulated to the brain.

🐦 Positive Gs refer to the upward acceleration that makes you feel heavier and presses you down in your chair. Pulling too many positive Gs can cause "gray-out," or a dimming of vision, and "black-out," which is unconsciousness.

🐦 Negative Gs refer to the downward acceleration that makes you feel lighter.

🐦 Negative Gs are also the reason pilots strap in so tightly: Pulling too many negative Gs can cause "red-out" because too much blood rushes to the head.

🐦 A red-out—which is essentially the inverse of a gray-out—can cause retinal eye damage and hemorrhagic strokes.

🐦 To prepare astronauts for spaceflight, NASA developed a specialized airplane known as the "vomit comet" that would fly parabolic arcs, providing astronauts a preview experience of weightlessness, which is zero Gs. The reason for the moniker "vomit comet" should be self-explanatory. Yuck.

🐦 The maximum Gs experienced during a space shuttle launch and re-entry is three Gs. Accelerations beyond 100 Gs, even if only momentary, are almost always fatal.

Giant Clams

✽ Giant clams (*Tridacna gigas*) are the largest living bivalve mollusks, and they usually live for 100 years or more.

✽ Discovered in 1817 off the coast of Sumatra, the largest known giant clam weighed 510 pounds (230 kg) without the living clam inside and measured 4½ feet (1.4 m) wide. Had the clam been alive, he or she would've weighed roughly 550 pounds (250 kg).

✽ Giant clams are incapable of locomotion—they can't make themselves move. This is called being "sessile." However, when in their larvae stage, they're capable of drifting. This is called being "planktonic."

✽ To reproduce, giant clams do a very cool thing: In any given area, all of the giant clams spawn at the exact same time!

G

✽ And since giant clams have both male and female reproductive organs (hence, my earlier "he and she" designation), they actually expel both eggs and sperm into the water simultaneously, which exponentially increases the chances of successful fertilization … and temporarily fog up the surrounding water a lot!

✽ In addition to filter feeding, giant clams rely on their symbiotic relationship with zooxanthellae to fulfill their nutritional and other needs. [*See* Zooxanthellae, *page 304.*]

Ginkgo Biloba

🍂 Pronounced *GIN-ko by-LO-bah* (with a "hard" g, as in "grasp"), most people think it's an herb, but it's not. It's actually the leaves from the maidenhair tree—also known as the ginkgo biloba tree.

🍂 The ginkgo tree is considered a living fossil. Evidence of the first appearance of the ginkgo tree dates back to the Early Jurassic period!

🍂 There are thriving ginkgo trees adorning the ground of temples in China that are believed to be over 1,500 years old!

🍂 Ginkgo trees are remarkably tolerant of air pollution, corrosive salts that are used on roadways, and are famous for their resistance to disease, pests, and fires. (They'd have to be to have survived millions of years!)

🍂 In the Western world, the ginkgo leaves are the most valued, whereas the seeds are considered far more medicinally important in traditional Chinese medicine.

🍂 Commonly referred to as simply "ginkgo," numerous studies on the possible health effects of ginkgo have been conducted for a variety of conditions.

🍂 Clinical data suggests that ginkgo is useful for treating a variety of ailments, including cerebrovascular insufficiency, which is a lack of blood flow to the brain accompanied by headaches, and peripheral arterial occlusive diseases such as Raynaud's disease. Ginkgo may also help prevent thrombus (blood clots) and is an overall good antioxidant.

꩜ To researchers' (and everyone's) dismay, taking ginkgo was found to be ineffective in lowering the overall incidence of dementia and Alzheimer's disease in the elderly, and ineffective in slowing cognitive decline, lowering blood pressure, or reducing the incidence of hypertension.

꩜ Some small studies of ginkgo for memory enhancement have shown some promising results. One study found significant improvement in attention capabilities in healthy individuals. Other small studies have found modest improvements for cognition in dementia patients, improving intermittent claudication (limping) symptoms. There is also conflicting evidence on the efficacy of ginkgo for treating tinnitus, which is ringing in the ears.

Gold

❧ The approximate total amount of gold mined in all of human history is 5.3 billion troy ounces, which is roughly equivalent to 182,000 tons, or 363,762,732 pounds.

❧ The chemical symbol for gold—one of the noble metals—is AU. This comes from the Latin word *aurum*, meaning "gold."

❧ Some of the many reasons that gold is used in electronics like computers is because it's a good conductor of electricity, and it is unaffected by air, moisture, and most corrosive reagents.

❧ The electrical conductivity of gold is 71 times that of copper.

❧ Because of its ability to reflect electromagnet radiation such as infrared, gold is used for the suits and helmets of astronauts and as a protective coating on manmade satellites.

❧ While there are measurable amounts of gold in the sea—approximately 0.004 parts per million—to date, no one has economically successfully extracted gold from sea water.

❧ A very dense metal, gold is 1.5 times denser than lead and 19.3 times denser than water.

❧ Sometimes used as embroidery thread, a single ounce of gold—about 28 g—can be stretched into a wire more than 5 miles (8 km) long.

G

🔖 In fact, gold is so incredibly malleable that a thread drawn from one ton of gold can stretch over 480,000 miles (772,000 km). That's comparable to stretching from Earth to the Moon and back again!

🔖 Only 24-karat gold, which is 100% pure, has no other added metals.

🔖 Eighteen-karat gold contains 75% pure gold, which is 750 parts gold per thousand parts.

🔖 Fourteen-karat gold is 58% pure gold, which is 585 parts per thousand. In fine jewelry, the remaining 42% is usually silver.

🔖 Ten-karat gold—which is often used for men's wedding rings—is 41.7% pure gold, which is 417 parts pure gold per thousand parts.

Great Lakes, Africa

🐦 Lake Malawi, Lake Tanganyika, and Lake Victoria are collectively known as the African Great Lakes.

🐦 Located in the East African Rift vicinity, these lakes are among the largest freshwater lakes in the world. Only 3% of the water on Earth is fresh water!

🐦 Formed between one and two million years ago, Lake Malawi is the smallest of the three, but it's still bigger than the country of Wales!

🐦 There are more fish species in Lake Malawi than any other lake. There are 850 different types of cichlids alone.

🐦 Lake Tanganyika is the longest freshwater lake in the world and the second deepest after Lake Baikal in Russia. [*See* Lake Baikal, *page 124.*]

🐦 Formed about three million years ago and fed by at least 50 inlets and streams, Lake Tanganyika's only outflow is the Lukuga River, which it feeds only during years of extremely high rainfall!

🐦 Ninety-eight percent of the lake's cichlids—which comprise two thirds of all the lake's fish—are unique to Tanganyika.

🐦 Formed about 400,000 years ago, Lake Victoria is the second-largest freshwater lake in the world. It is also the source of the longest branch of the Nile River.

G

⚬ Since its formation, Lake Victoria has completely dried up at least three times, the most recent being about 17,300 years ago. It began refilling about 14,700 years ago.

⚬ While huge in surface at 26,254 square miles (68,800 km²), Lake Victoria is quite shallow, with an average depth of only 130 feet (40 m).

⚬ Lake Victoria is facing a bevy of environmental problems, including pollution, oxygen depletion, invasive species introduction, overfishing, and algae growth.

⚬ Lake Victoria has lost 80% of its indigenous fish species. Rehabilitation projects such as The Lake Victoria Environmental Management Project are actively identifying, analyzing, prioritizing, and pursuing corrective measures to reverse this.

Group Names

- ❧ A group of gorillas is called a "band."
- ❧ A group of crows is called a "murder."
- ❧ A group of feral cats is called a "clowder."
- ❧ A group of kangaroos is called a "mob" or "troop."
- ❧ A group of lions is called a "pride."
- ❧ A group of whales is (most commonly) called a "pod."
- ❧ A group of geese is called a "gaggle."

Group Names
(Less Commonly Used Names)

❧ Published in the 1486, *The Book of St Albans* contains essays on "gentlemanly pursuits," such as falconry, hunting, and heraldry. One of the sections describes group names of animals in rather poetic fashion. And while these fanciful terms are less commonly used, they are cool nonetheless:

- ❧ A group of leopards is called a "leap."
- ❧ A group of jellyfish is called a "smack."
- ❧ A group of peacocks is called an "ostentation."
- ❧ A group of ferrets is called a "business."
- ❧ A group of larks is called an "exaltation."
- ❧ A group of owls is called a "parliament."
- ❧ A group of foxes is called a "skulk."

Hair

🍂 All mammals have hair, even marine mammals like dolphins.

🍂 Although there are very few, dolphins' hairs fall out shortly after birth.

🍂 The average human has 100,000 to 150,000 hair follicles on their scalp.

🍂 The average human loses between 50 to 100 hairs per day.

🍂 Human hair grows autonomously, meaning each hair is on its own individual cycle. If all our hair were on the same cycle, we'd molt!

🍂 Human hair has the highest rate of cell division—called mitosis. On average, human hair grows about 0.011 inches (0.3 mm) a day and 0.4 inches (1 cm) per month.

H

🍂 Xie Qiuping, a Chinese woman whose hair measures about 18 feet, 5.54 inches long (about 5.67 m), has the longest human hair in the world!

🍂 Hair can serve a wide range of purposes in the animal kingdom. It can act as camouflage, provide heat regulation, communicate mating or warning signals, and sometimes even offer defensive or offensive protection as it does for porcupines, whose quills are modified hairs.

Head-Bobbing Birds

♾ Head bobbing when walking is a behavior unique to birds.

♾ At least eight of the 27 families of birds, including quails, cranes, chickens, and magpies, bob their heads while they walk. So, what's up with that?

♾ Common theories posit that the head-bobbing action of birds assists with balance, provides depth perception, and sharpens vision.

PROPER MOTIVATION

 However, the studies with the most compelling evidence are those suggesting that head-bobbing behaviors function to stabilize birds' visual surroundings. Want some compelling evidence? A 1978 study specifically researched whether a pigeon walking on a treadmill (yes, you read that correctly) would still bob his head. Why a treadmill? The speed, stability, and visual surroundings would remain relatively the same for the pigeon, which is the basis for the theory. The result? When walking on a treadmill, the pigeons no longer bobbed their heads! Is this conclusive proof? No, more research is necessary. But it's a really cool study.

Health and Music

"Music has charms to soothe a savage breast."
—William Congreve, 1697

H

❧ Hospital operating room studies have found that listening to music can prevent distraction, minimize annoyance, reduce stress, and diminish the anxiety of both patients and staff alike.

❧ Scientific research has demonstrated that listening to music can reduce patients' sensation of pain.

❧ Both MRI (magnetic resonance imaging) and PET (positron emission tomography) scans reveal that the pleasure, reward, emotion, and arousal regions of our brain light up when listening to music we love.

❧ Singing sentences (instead of speaking them) has been shown to improve communication capabilities between caregivers and Alzheimer's patients.

❧ Clinical studies have demonstrated that music can improve the precision of fine motor skills, walking, and posture control in people affected by Parkinson's or Alzheimer's, multiple sclerosis, and ataxia.

❧ When people are listening to music they find immensely pleasurable, the neurotransmitter dopamine—the feel-good, happy chemical—is released into our brains.

❧ Music has been shown to be an effective tool in lessening depression, as part of a multiprong effort.

⚬ Contrary to popular belief, the type of music that's most effective for improving physical and emotional outcomes isn't necessarily Mozart or Bach. Whatever type of music speaks to you or your loved one is the kind that works best—be that rock 'n' roll, reggae, jazz, hip hop, country, blues, pop, gospel, bluegrass, disco, Motown, folk, funk, or classical!

Heart

- Electricity going through your heart makes the muscle cells contract.

- The study of the human heart and its various disorders is called cardiology.

- Every day, the average adult heart pumps 4,000 gallons (15,140 liters) of blood.

- Every minute, the average heart pumps 5.28 quarts (5 liters) of blood.

- On average, an adult heart beats 30 million times per year.

H

- The *lub-dub* sound your heart makes is the vibration in your blood and tissues caused by the heart valves slamming shut. This is a very important process as this prevents your blood from flowing in the wrong direction—like backward!

Hippocampus

❧ Located in your brain, the hippocampus has a striking resemblance to a sea horse. The medical term "hippocampus" comes from the Greek word *hippos*—meaning "horse"—and *kampos*—meaning "sea monster."

❧ Essential to your ability to transfer immediate or short-term memories into long-term memories, your hippocampus is a vital part of the brain. Recognizing that you ate breakfast ten minutes go, walked the family pet, and asked your boss for a raise has to do with your ability to acquire, store, and retrieve memories.

❧ The hippocampus is also essential to both your spatial memory and your spatial navigation. So, if you need directions to get to places you've driven to countless times before, and if you can't read a map to save your life, then your hippocampus may be partially to blame.

H

❧ Ever wonder why certain smells can trigger a memory? Your hippocampus is part of the highly interconnected limbic system, your "primitive" brain, which includes the amygdala. The amygdala processes memory and emotion, and your olfactory cortex is involved in your sense of smell.

❧ Known for their extraordinary navigation skills—and thus the focus of many scientific, peer-reviewed studies—London taxi drivers have greater gray matter volume in their right posterior hippocampus than non-taxi drivers, London bus drivers, and medical doctors. And the more years on the job, the greater that volume!

Horses

꙲ There's no such thing as a white horse. They're actually all called gray horses (by those in the know) because they have little black and white hairs that combine to make them look white.

꙲ Unlike many animals, horses aren't colorblind.

꙲ Horses' teeth never stop growing.

꙲ The American Quarter Horse is one of the fastest land animals in the world, averaging a speed of 48 mph (77 km/h)! For some perspective, a coyote averages 43 mph (69 km/h), and a lion averages 51 mph (82 km/h).

H ꙲ On average, an adult horse's intestine is 89 feet (27 m) long!

꙲ James Watt—for whom the watt unit of power was named—popularized the term "horsepower" as a marketing tool to promote his steam engine. The term seems to have been originally coined by Thomas Savery in his 1702 book *The Minor's Friend*, explaining the benefits of the steam engine over horses.

꙲ Horses cannot vomit, because they have no gag reflex.

꙲ Horses can't breathe through their mouths.

꙲ Horses have seven blood types.

꙲ Horse meat is considered a delicacy in many countries outside the U.S. When it's served raw in Japan, it is called "cherry blossom."

Human Hands

❧ The scientific name for the back of the hand is "opisthenar."

❧ The technical term for fingerprint, also known as a dermal ridge, is "dactylogram." The primary function of fingerprints is to provide friction and traction so the hand can grasp and hold onto objects.

❧ Contrary to popular belief, *Homo sapiens* are not the only species who have opposable thumbs; monkeys and chimpanzees easily move their thumb into a position opposite to their other four fingers.

❧ What people *can* do that no other creature in the animal kingdom can is "ulnar opposition"—the ability to bring the small finger and ring finger across the palm to meet the thumb. This provides superior gripping and grasping capabilities.

❧ The human hand has approximately 27 bones—that's more than half of all the bones in your entire upper limb.

❧ There are no muscles in your fingers and thumbs, only tendons, which are connected to 17 muscles in your palm and 18 muscles in your forearm. The movement of your fingers and thumb is like pulling a marionette's strings.

❧ Each human hand has 29 major joints (give or take a few), 123 ligaments, 48 nerves, and 30 arteries.

❧ Phalanges are the bones in the fingers and toes. The word "phalanges" comes from the Latin word for "row of soldiers."

H

Hurricanes

❧ In the Northeast Pacific Basin and in the Atlantic, a strong tropical cyclone is called a hurricane.

❧ In the Northwestern Pacific, the same storm would be called a typhoon.

❧ In the Southern Hemisphere and the Indian Ocean, it would simply be called a cyclone.

❧ A tropical cyclone has counterclockwise wind flow in the Northern Hemisphere and clockwise wind flow in the Southern Hemisphere because of the Coriolis effect. [*See* Coriolis Effect, *page 34.*]

H

❧ The power of a cyclone comes from the heat of condensation of warm, moist air pulled up into the higher atmosphere.

❧ Around the world there are six regional specialized meteorological centers designated by the World Meteorological Organization that are responsible for tracking, classifying, and naming storms in their regions.

❧ In the United States, hurricane force is classified on the Saffir–Simpson Hurricane Wind Scale (SSHWS) on a five-point scale: category one to five. Other regions around the world use other classification schemes and nomenclature.

    ❧ Tropical cyclones are sometimes compared to each other based upon the lowest barometric pressure measured at the eye of the storm.

    ❧ The lower the barometric pressure, the stronger the storm. The strongest storm ever measured was Typhoon Tip in the Western North Pacific Ocean basin at 25.7 inches of mercury (900 hPa) in 1979. Note: Inches of mercury is a common unit of measure for barometric pressure used by meteorologists.

Hydrogen

⤳ Hydrogen is the simplest element.

⤳ Each atom of hydrogen has only one proton.

⤳ Hydrogen is the lightest element and is a gas at standard ambient temperature (77°F/25°C) and pressure 100 kPa (kilopascals).

⤳ Hydrogen condenses to a liquid at temperatures of -423°F (-253°C).

⤳ Hydrogen is the most plentiful gas in the universe.

⤳ Hydrogen has the highest energy content of any common fuel by weight—about three times more than gasoline—but the lowest energy content by volume—about four times less than gasoline.

⤳ Hydrogen can be produced from a variety of resources, including water, fossil fuels, or biomass, and is a byproduct of other chemical processes.

⤳ Hydrogen is found in all growing things.

⤳ Hydrogen is an abundant element in Earth's crust.

⤳ Stars like the Sun are made primarily of hydrogen.

H

International Morse Code

🕊 A "code" replaces words, phrases, or sentences with groups of letters, symbols, sounds, or numbers. A cipher, on the other hand, rearranges letters or uses substitutes to disguise a message.

🕊 In 1835, while a professor of arts and design at New York University, Samuel Morse used an electromagnet to produce written code on strips of paper, thereby inventing Morse code. One year later, he modified the code to a system of dots and dashes (• –), which is still used today.

🕊 Congress funded Morse to make a prototype that would transmit the code between Washington, DC, and Baltimore, Maryland.

🕊 The first message to be sent by Morse code was the Biblical verse "What hath God wrought?" [Numbers XXIII, 23] which was transmitted in 1844 over the first DC-to-Baltimore telegraph line (a distance of about 40 miles/65 km). The message was sent from the Supreme Court to Morse's colleague in Baltimore.

🕊 There are two other earlier versions of Morse code: American Morse code and the modified version used by German railways, both of which used dots and dashes. The international version is today's standard.

I

Particularly useful sending and receiving signals:

- AAAAA etc.: Call sign. I have a message.
- AAA: End of sentence. More follows.
- EEEEE etc.: Error. Start from last correct word.
- TTTTT etc.: I am receiving you.
- IMI: Repeat sign. I do not understand.
- K: I am ready. Start message.
- R: Message received.
- AR: End of message.

Particularly useful words to commit to memory:

- SOS: • • • | – – – | • • •
- HELP: • • • • | • | • – • • | • – – •
- LOST: • – • • | – – – | • • • | –
- INJURY: • • | – • | • – – – | • • – | • – • | – • – –

Internet

🌿 Often referred to as simply "the Net," the Internet is a short form of the technical term "internetwork."

🌿 The Internet is a network of networks, as opposed to the World Wide Web, which is one of the services of the Internet. [*See* World Wide Web, *page 272.*]

🌿 Bob Kahn and Vint Cerf—known as the fathers of the Internet—credit Al Gore as the first political leader who recognized the importance of the Internet.

🌿 Kahn and Cerf invented the Transmission Protocol (TCP) and the Internet Protocol (IP), which are the fundamental basis for Internet communication.

🌿 In the United States, the first commercial dial-up Internet Service Provider (ISP) was called "The World." Other early 1980s ISPs include Netcom, PSINet, and UUNET.

🌿 In January 2010, astronaut T. J. Creamer was the first person to post an unassisted tweet—a message to his Twitter account—from space, which he did from the International Space Station!

🌿 Founded in 1992, an international nonprofit organization called the Internet Society (ISOC) was established "to assure the open development, evolution, and use of the Internet for the benefit of all people throughout the world."

I

Invasive Species:
The Burmese Python

❧ The native habitat range of the Burmese python is southern and southeastern Asia. However, these snakes are an invasive species in Florida, where they have been released and are rapidly reproducing and consuming large numbers of mammals and birds, including federally endangered species, in places such as the Everglades National Park.

❧ Confined pythons exhibit impressive muscular strength and persistence, and they can sometimes escape even well-built cages. Some owners who felt unprepared to properly care for such large and voracious pets have also been known to intentionally release them.

I

❧ Climate- and habitat-matching models indicate that a substantial portion of the U.S. may be vulnerable to this ostensibly tropical species. So, they have the potential to spread far beyond Florida.

❧ Capable of eating meals that are equal to their own body weight, the Burmese python is a significant risk to a large number of imperiled mammal and bird species.

❧ One would think that large prey such as herons, with their incredibly powerful, sharp beaks, could easily kill a python by puncturing their gut, as they are well known to do. However, pythons are able to encase sharp bird beaks in their connective tissue, where they remain inside the body cavity for the lifespan of the snake!

IT'S ALL A MATTER OF PERSPECTIVE

Iodine

⚬ Iodine is an element essential for health and derives from the Greek word meaning "violet" or "purple," which refers to the color of the elemental iodine vapor.

⚬ Iodine is one of the earliest elements whose radioisotopes were used in what's now referred to as nuclear medicine.

⚬ Iodine 131, which has 53 protons and 78 neutrons, has too many neutrons in its nucleus, making it unstable and radioactive.

⚬ Iodine 131 became one of the earliest radioactive tracers, which is the very basis of a variety of valuable medical imaging systems such as PET (positron emission tomography) scans.

⚬ The most common, stable form of iodine has an atomic number of 53 protons and an atomic weight of 127 (53 protons, plus 74 neutrons).

⚬ Dietary iodine deficiency affects about two billion people worldwide.

⚬ Dietary iodine deficiency is the leading preventable cause of intellectual disabilities. Primarily seen in countries that don't use iodized salt, insufficient iodine in one's diet can lead to a goiter—a swelling of the thyroid gland.

Ionosphere

❧ The ionosphere is the top layer of the atmosphere. The other layers are the troposphere, where weather happens; the stratosphere, where the ozone layer is located; and the mesosphere, where most meteors burn up upon entry.

❧ The ionosphere extends from about 50 to 250 miles (80 to 400 km) in height from Earth's surface.

❧ The Northern Lights (also known as "aurora borealis") and Southern Lights (also known as "aurora australis") form in the ionosphere.

❧ The density of the atmosphere in the ionosphere is very low. So, when high-energy solar radiation ionizes a gas molecule by knocking an electron off, it can take some time before the positively charged ion and the negatively charged electron find each other and recombine to form a neutral gas molecule. That's why it's called the *ion*osphere!

❧ The ionosphere consists of at least three major layers—D, E, and F— with some variations when the F layer splits into two distinct layers, F1 and F2, or when a phenomenon called a "sporadic E layer" forms, which can't be predicted, hence the term.

❧ High-frequency shortwave radio signals are refracted by various layers of the ionosphere, which allow them to reach great distances over the horizon. How far? That depends on a variety of atmospheric conditions. This is the kind of information that HAM radio folks really get excited about!

IQ

❧ IQ stands for intelligence quotient.

❧ The study of the heritability of IQ—the investigation as to whether or not one can "inherit" intelligence—has been going on since the 19th century.

❧ The heritability of IQ is a hotly contested topic in the academic and scientific communities. The primary points of contention include if and to what extent one's IQ is the result of "nature" (inherited) or "nurture" (the environment, which includes how one is brought up, etc.).

❧ The theory of multiple intelligences posits that there are eight areas of intelligence that we learn or acquire: Interpersonal, intrapersonal, bodily kinesthetic, linguistic, logical-mathematical, musical, naturalistic, and spatial-visual.

❧ A standard IQ test isn't so much a definition of one's intelligence. Rather, it's a test that measures a person's cognitive ability compared to the population at large, and purportedly measures a child's "potential." Thus, one could say that intelligence *is* what the intelligence test measures ... which may or may not be predictive of anything.

✦ Albert Einstein didn't speak until he was four years old. That would have boded poorly for his "potential," had he been tested … and we all know how HE turned out!

✦ Recent studies have some scientists suggesting that "bigger brains are smarter" supports the argument that there's a genetic connection between brain size and IQ.

Jacobson's Organ

❧ We tend to think that people have five senses that perform the basic functions of seeing, hearing, tasting, touching, and smelling. However, we also have a specialized organ in our noses that can perform two more rather amazing tasks, namely: Detecting trace quantities of chemicals, and recognizing subtle human-to-human communication. This subtle communication is emitted via pheromones, which are chemical messengers that carry information between individuals of the same species.

❧ Jacobson's organ—or the "vomeronasal organ" or "vomeronasal pit"— is an olfactory sense organ first discovered in humans by Frederik Ruysch in 1703. It is named for Ludwig Jacobson who described it in 1813.

❧ Where in the nose is this organ? The telltale paired bilateral vomeronasal pits are located in the front third of the nasal septum, the bone that divides your nostrils.

❧ **How we smell stuff 101:** When we detect an odor, the olfactory bulb sends neuronal signals to your neocortex and limbic system. This system is involved in memory and emotion, particularly reflexive emotions, such as fear. Your neocortex is involved with most everything to do with cognition, such as taste, touch, sight, sound, smell, balance, temperature, spatial perceptions, emotional responses, and your ability to communicate.

❧ Vomeronasal systems have long been known to exist in all fetal humans!

❧ Based upon endoscopy—fiber-optic scoping—not all human adults have a Jacobson's organ and/or a functioning vomeronasal system.

Jazz

🔹 Hospital operating room studies have found that listening to music can minimize annoyance and diminish anxiety. One study that polled doctors, nurses, and patients found that 71.6% of the respondents preferred listening to jazz over other types of music in the operating room. [*See* Health and Music, *page 80.*]

🔹 When professional jazz musicians switch from playing songs they regularly play to playing improvisational music, MRI scans reveal that their prefrontal cortex goes silent.

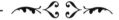

❧ What does the brain's prefrontal cortex do? Well, it does lots of things, but it's known for being in charge of conscious self-monitoring, evaluating, and "correcting" what one's doing. In other words, when the prefrontal cortex goes silent, inhibitions are muted.

❧ Intensely pleasurable responses to jazz music, which happens in the pleasure, reward, and emotion regions of the brain, can be clearly seen on functional MRI scans.

❧ In one study using a PET scan, scientists found that when people listened to music they absolutely love—and specifically, it had to be music that literally gave them "the chills" or "chills down their spine," which for some people is jazz—there was an increased blood flow to the regions in the brain involved in reward/motivation, emotion, and arousal.

❧ As compared to other musicians and non-musicians, research suggests that jazz musicians have greater amplitude, or sensitivity to sound waves, across all sound features, which suggests that jazz musicians' perceptual skills and brain processing may be influencing their music.

❧ In 1920s America, critics of jazz music said it caused mental and physical impairments. Proponents of jazz responded in kind, asserting that jazz didn't cause impairments, it cured them!

Jellyfish

🐦 Despites its moniker, jellyfish aren't fish at all—they're members of the phylum Cnidaria.

🐦 Having roamed the seas 500 to 700 million years, jellyfish are believed to be the oldest multi-organ animal.

🐦 Pouring fresh water on a jellyfish sting that occurred in salt water is ill advised. Doing so can actually change the osmotic pressure, and more venom can be released. To the best of our knowledge, jellyfish don't have brains.

🐦 Jellyfish are comprised of over 95% water.

🐦 A group of jellyfish can be called a "smack," "bloom," or "swarm."

J

🐦 The ocean sunfish, which is the heaviest known bony fish in the world, primarily dine on jellyfish and are considered a major jellyfish predator.

🐦 Reproducing both sexually and asexually, jellyfish are usually male or female, although there are some species that are hermaphrodites and have both male and female reproductive organs.

Joint Cracking

- Besides cracking one's knuckles, joints that can make sounds in the human body include the knees, ankles, back, and neck.

- Whether the joints are making popping, snapping, grinding, or cracking sounds, there are scientific reasons that explain why they make these noises (that probably annoy your mom):

- **Escaping gases:**

  - Inside your joints is synovial fluid, which acts as a lubricant. That fluid contains the gases oxygen, nitrogen, and carbon dioxide.
  - When you pop or crack a joint, you stretch the joint capsule. Gas is rapidly released, which forms bubbles.
  - In order to crack the same knuckle again, you have to wait until the gases return to the synovial fluid.

- **Movement of joints, tendons, and ligaments:**

  - When a joint moves, the tendon's position changes and moves slightly out of place.
  - You may hear a snapping sound as the tendon returns to its original position.
  - In addition, your ligaments may tighten as you move your joints.
  - This commonly occurs in knees or ankles and can make a cracking sound.

🍂 **Rough surfaces:**

🍃 Arthritic joints make sounds caused by the loss of smooth cartilage and the roughness of the joint surface.

🍃 Healthcare professionals advise that if you're feeling pain when your joints pop or crack, seek medical attention.

🍂 In terms of knuckle cracking, some studies show that knuckle cracking doesn't cause serious harm. Other studies show that repetitive knuckle cracking can do some damage to the soft tissue of the joint. Also, knuckle cracking may lead to a weak grip and a swollen hand.

🍂 Cracking my knuckles makes my hands feel better. But perhaps I should consider stopping and talk to my doctor!

Joints

⚜ The word "joint" is from the Latin word *jungere* meaning "to join." Whether it's two bones or multiple bones, joints are formed wherever bones connect to other bones.

⚜ Not all joints move: Your skull is an immovable joint, but it wasn't always immovable. At birth, humans have eight separate cranial bones (that thankfully squish together during the birthing process). Over time, those bones fuse together to form a single, protective helmet for the brain.

⚜ There are over 230 moveable and semimoveable joints in the average adult.

⚜ Each human hand has 29 major joints (give or take a few), and each foot has 33.

⚜ The classification of joints is determined by how much articulation they have, or by how much and to what extent they can move.

⚜ The most maneuverable type of joint is the ball and socket configuration: the shoulder joint can move 350 degrees. Then, continuing in descending order are saddle joints, ellipsoidal joints, pivot joints, hinge joints, and gliding joints.

⚜ The knee joint is the largest hinged joint in the body.

⚜ In healthy joints, the bones are encased in smooth cartilage. Cartilage is 65–80% water.

🐾 Osteoarthritis is caused when the cartilage in a joint breaks down and wears away.

🐾 Rheumatoid arthritis is an autoimmune disease that is caused when the immune system attacks healthy tissues in the joints as if they were disease-causing germs.

Jumping Genes

🐦 Initially referred to as "jumping genes," transposons are bits of DNA capable of moving from one place in the genome to another.

🐦 Barbara McClintock won the 1983 Nobel Laureate in Physiology or Medicine for her 1948 discovery of these mobile genetic elements.

🐦 Dr. McClintock performed this and other groundbreaking genetics work by studying the chromosomal structure and inheritance patterns in the brightly colored kernels of flint corn (often called Indian corn).

🐦 There are two main types of transposons, classified by how they "jump" around the genome:

J

    🐦 During the movement of Class I (or retrotransposons), they go from DNA to RNA and back to DNA again, in a "copy and paste" process, leaving the original copy behind.

    🐦 During the movement of Class II (or DNA transposons), the jumping genes undergo what's sometimes described as a "cut and paste" process, where they cut themselves out of the chromosome and move to another location.

🐦 More than half of the DNA sequence is made up of various types of transposable elements, while only about 1.5% of our DNA encodes our genes' protein-coding regions (exons).

🐦 It's believed that transposons play an important role in evolution, in part, by "shuffling the deck" of genes and chromosomes into new arrangements!

Jupiter

🐦 The largest planet in our solar system is Jupiter. Known for its Giant Red Spot, Jupiter is the fifth planet from the Sun and has a diameter of 88,736 miles (142,800 km). Earth's is 7,926 miles (12,753 km), so Earth could fit inside Jupiter over 1,000 times.

🐦 Jupiter's Giant Red Spot—which is actually a gigantic storm that rotates counterclockwise—is 24,860 miles (40,000 km) wide! This storm has been raging for somewhere between 182 and 347 years ... or more.

🐦 Jupiter is about 508 million miles (817 million km) from the Sun, and its revolution around the Sun takes 12 Earth years.

🐦 Jupiter has three very faint thin rings: Gossamer, Halo, and Main.

🐦 Jupiter has 63 known moons—technically only 50 are official.

🐦 Made mostly of hydrogen, and to a lesser extent helium, Jupiter has the largest planetary atmosphere in our solar system, which, from lowest to highest layer, includes a troposphere, stratosphere, thermosphere, and exosphere!

🐦 A helium-filled balloon would sink on Jupiter because helium is heavier than hydrogen.

🐦 The beautiful alternating light and dark bands of Jupiter are called "belts" and "zones." The belts are the darker ones; the zones are the lighter ones.

🐦 Jupiter has no solid surfaces, so it's not possible to land on Jupiter!

J

Jurassic Park
"Easter Egg"

❧ The "dinosaur DNA" sequence in Michael Crichton's book *Jurassic Park* wasn't even plausibly close to a dinosaur. It was actually the DNA of an E. coli plasmid, which is a circle of DNA that replicate in bacteria.

❧ A brilliant National Institutes of Health (NIH) scientist named Dr. Mark Boguski identified this scientific faux pas and actually wrote to Mr. Crichton about this shortcoming. Dr. Mark Boguski suggested to Mr. Crichton that a bird-based DNA would be more scientifically accurate.

❧ By the personal request of Michael Crichton, the "dinosaur DNA" sequence in his sequel, *The Lost World*, was designed by Dr. Mark Boguski.

❧ Dr. Mark Boguski inserted a *secret message* into his bird-based (chicken) and "dinosaur" DNA.

❧ If you translate his combination dino/chick sequence into its encoded protein, and compare it to the original chicken sequence, the inserted non-chicken amino acids spell out, "MARK WAS HERE NIH!"

❧ The term "Easter egg" refers to how we geeks and nerds refer to secret messages like Dr. Boguski's insertion; hidden signature motifs, like Alfred Hitchcock making secret cameo appearances in his own films; and hidden functions like Microsoft Excel 97's hidden flight simulator.

To access Microsoft Excel 97's Easter egg in the hidden flight simulator program (which also includes a very clever credit roll), you'd do the following: On a new worksheet, press F5; type X97:L97 and hit Enter; press the tab key; hold Ctrl-Shift; click on the chart wizard toolbar button; use the mouse to fly around. The right button is forward, and the left button is reverse.

JURASSIC PARK EASTER EGGS

K
(a.k.a. Potassium)

🍂 In 1807, a Cornish chemist named Sir Humphry Davy attempted something no one else had ever considered trying before: He used electricity, a newly discovered force, and revealed a brand-new element! Using a voltaic pile, the first electric battery invented by Alessandro Volta in 1800, on the caustic, molten salt potash, Davy derived the previously unknown element potassium! And thus the field of electrochemistry was born. (Fun fact: The word "volt" was coined for Alessandro Volta.)

🍂 Incidentally, Davy first made a name for himself back in 1800 for preparing and inhaling nitrous oxide (laughing gas) and publishing his findings in *Researches, Chemical and Philosophical*. The following year he was hired as a chemistry lecturer at the Royal Institution. But getting back to potassium …

🍂 Because potassium is very electropositive, it reacts so violently with water that it causes eruptions! Why does this happen? Read on!

🍂 When combined with water, potassium forms the colorless compound of potassium hydroxide solution and hydrogen gas, which creates a huge amount of heat and burns with a brilliant purple/blue flame!

🍂 When the released hydrogen comes in contact with oxygen, it will ignite! That's why it must be stored the same way we store gasoline (petrol) and kerosene.

🐦 It should come as no surprise that potassium nitrate (a compound of potassium) is used in fireworks and gunpowder!

🐦 Potassium is insoluble in water, yet it's such a soft solid that it can be easily cut with a knife.

🐦 Essential for the proper function of all cells, tissues, and organs in the human body, potassium is crucial to healthy heart function, plays a key role in skeletal and smooth muscle contraction—such as normal digestive and muscular function—helps maintain normal nervous system and fluid balance, and helps remove toxins from the tissues.

🐦 Eating foods rich in potassium can help counterbalance diets too high in salt. While people commonly think of bananas as the "go-to" food for potassium, there are many foods that have a far higher content. A medium banana has about 400 mg of potassium. However, check these out: Just ½ cup of raisins has a whopping 1,125 mg of potassium; 1 cup of lima beans has 1,000 mg; ½ avocado has 742 mg; one baked sweet potato has 690 mg; and one baked potato is 900 mg. Just saying!

🐦 So, why do we use a "K" as the symbol for potassium? It comes from the Latin word *kalium*, meaning "pot ashes," and the Arabic word *qali*, meaning "alkali." The English word "potassium" comes from the word "potash," and the suffix "-ium," meaning "bigger" or "more complicated than."

K2

～ Part of the Karakoram Range located in South Asia, K2 is the second-highest mountain in the world; the tallest mountain is Mount Everest.

～ K2 is an impressive 28,251 feet (8,611 m) at its highest point of elevation. The air is so thin at the summit that climbers can only inhale about one-third of the oxygen they would normally get from a breath at sea level.

～ K2 is pyramid-shaped with devastatingly sheer slopes that drop precipitously.

～ The unusual name "K2" is attributed to Thomas Montgomerie, a member of the 1856 European survey team, who conducted the very first survey of the Karakoram Range. Montgomerie had sketched the two most prominent peaks from his vantage point, standing on Mt. Haramukh 125 miles (201 km) to the south. He labeled them simply "K1" and "K2": The "K" for Karakoram and the "2" to designate it as the second peak in the range.

～ K2 also goes by a number of other names, including Mountaineer's Mountain, Mount Godwin-Austen, Savage Mountain, Balti: Chogori and Sarikoli: Mount Qogir.

～ Were the other peaks originally designated K1, K3, K4, and K5? Yes! They've since been named Masherbrum, Gasherbrum IV, Gasherbrum II, and Gasherbrum I, respectively.

～ Tragically, one in five people who attempt to reach the summit of K2 dies while trying.

Kakapos

❧ Considered effectively extinct for three-quarters of the 20th century, this New Zealand flightless bird is making a comeback! The Kakapo (*Strigops habroptilus*) is one incredibly unique bird—that's fabulous news!

❧ Why was the Kakapo "effectively" extinct? Well, in 1974 there was only one known kakapo—an aging male—named Jonathon Livingston, but since kakapos live between 80 and 100 years, they weren't extinct yet.

❧ Kakapos are the only known flightless parrot. Yet, you're likely to see them perched high in trees! How do they get there? Skilled climbers, they use their wings for balance and their beak and strong claws to pull and grip their way up and down trees and vines. An amazing thing to witness!

❧ Kakapos are the only nocturnal parrot.

❧ The kakapo is the only parrot with an inflatable thoracic air sac, which he uses as part of the mating ritual. He can puff up his entire body to the size of a basketball in order to impress the females!

❧ Kakapos are the world's heaviest parrot, with males weighing nearly 9 pounds (4 kg)! Females usually weigh about one-third less than their male counterparts.

❧ Although categorized as a parrot and belonging to the *Strigopidae* family of New Zealand parrots, kakapos are really not that biologically close to other parrots. In fact, kakapos have some rather owl-like features, such as furlike disks around its eyes and soft feathers.

K

Kiwi Birds

❧ The kiwi bird is often thought of as native to New Zealand, but DNA evidence strongly suggests that its ancestors arrived there via Australia.

❧ One could argue that the kiwi is the closest thing to a mammal in the bird world. Consider this:

> ～ For starters, rather than having air-filled hollow bones as birds do, kiwi birds have bone marrow.
>
> ～ The kiwi bird is the only bird in the world with nostrils at the end of its bill. In fact, kiwi birds' olfactory bulb—responsible for odor perception and transmission—is structured more closely to that of mammals, so their sense of smell is superior to that of most birds.
>
> ～ Kiwi birds have long whiskers, large ear openings, and their feathers look more like fuzzy hair. [*See* Kiwi Fruit, *page 115.*]
>
> ～ The blood temperature of kiwi birds is nearly the same as mammals.

K

❧ A female kiwi bird's egg can weigh up to one-fifth of her body weight. The largest recorded egg weighed 17.6 ounces—more than a pint (500 g), or the volume of six hen's eggs!

❧ Usually monogamous, kiwi birds mate for life.

❧ On average only 5% of kiwi chicks survive to adulthood.

❧ Kiwis are flightless birds of which there are five recognized species, including the great spotted kiwi, the little spotted kiwi, the okarito kiwi, the southern brown kiwi, and the North Island brown kiwi—all are endangered.

Kiwi Fruit

❧ If you've never eaten a kiwi: They're rather odd-looking, fuzzy, brownish-green, little, egg-shaped fruit. But inside, they have beautifully translucent, emerald green flesh with perfectly concentric circles of little black seeds—and they taste like an exotic cross between a strawberry, a pineapple, a melon, and a banana.

❧ Contrary to popular belief, kiwis were not native to Australia or New Zealand; they're actually originally from China. They also weren't called kiwis, either. They were called *Yang Tao*.

❧ Here's what happened: Missionaries brought kiwis to New Zealand, where they were renamed Chinese gooseberries.

K

❧ Then, in the early 1960s, some Americans "discovered" this wonderful little fruit and thought that they would be well suited for the American palette.

❧ ... But there was a marketing problem:

❧ They were called Chinese gooseberries. However, gooseberries were known to be prone to plant diseases—which could cause import problems. They were briefly renamed melonettes, but that didn't fly, because melons were subject to import tariffs. Finally, the New Zealand growers suggested "kiwi" to evoke their national symbol: the fuzzy little brown-coated bird— which also resembles the fruit!

❧ Kiwis are one of a handful of fruits that are green on the inside when ripe!

❧ And step aside orange: Ounce for ounce, no fruit provides you with more vitamin C than the kiwi, sans the guava.

❧ Similar to dark chocolate, eating kiwis regularly has been shown to reduce the "platelet aggregation response"—which means they may help lower your risk of developing blood clots.

Komodo Dragons

🐦 Komodo dragons are the largest living lizards in the world.

🐦 Found only on the islands of Komodo, Rinca, Gili Motang, and Flores, the Komodo dragon has the smallest territorial range of any of the world's large carnivores.

🐦 Within moments of hatching, newborn Komodo dragons have to immediately find a tree and climb high up or they'll be eaten by their cannibalistic elders. Yikes.

🐦 Komodo dragons are carnivorous creatures (which you've surely surmised!), and up to 10% of the adult Komodo diet consist of smaller Komodo dragons. They also enjoy boar, deer, water buffalo, civet cats, rats, birds, fish, snakes, chickens, goats, eggs, carrion … and the occasional human.

K

🐦 Komodo dragons can store enough fat in their tails to survive for one to one and a half months without eating or drinking—which, in times drought, is very handy!

🐦 In a single meal, Komodo dragons have been known to eat up to 80% of their body weight.

🐦 Don't get bitten by a Komodo. Their red saliva is jam-packed with such dangerous, virulent bacteria like E. coli that death by blood poisoning can happen in just one to five days. That is, if they don't eat you first.

Krebs Cycle

❧ Krebs cycle is one of our body's most important processes, and yet, there doesn't seem to be a simple explanation as to what it is and why it matters! So, here you go:

❧ Cellular respiration is simply the process by which we turn sugar from the foods we eat into usable energy within our cells. Without cellular respiration, we, the human race, would die.

❧ The Krebs cycle is the second of three stages necessary for cellular respiration; the other two stages are glycolysis (the first stage) and electron transport/oxidative phosphorylation (the third stage).

❧ In animals, the Krebs cycle occurs in the mitochondria, inside the cells; in plants, it occurs in the chloroplasts.

K

❧ Traveling via the bloodstream, glucose (sugar) is what stokes the mitochondrial furnaces responsible for your brainpower. And glucose is the only fuel our brain knows how to use!

❧ And so there's no confusion, the Krebs cycle may also find it referred to as both the tricarboxylic acid cycle (TCA) and the citric acid cycle.

❧ Finally, to give credit where credit is due, Sir Hans A. Krebs—for which the process is named—proposed the concept in 1937, and he was awarded the Nobel Prize in Medicine in 1953 for his outstanding contribution to medicine.

Krypton

◆ Generally chemically unreactive, as all noble gases, krypton is chemical element number 36 on the periodic table.

◆ Through the process of distilling liquid air, Sir William Ramsay and Morris Travers discovered krypton in 1898.

◆ In fact, Sir William Ramsay and Morris Travers discovered three new noble gases in this manner: Krypton, from the Greek word *kryptos*, meaning "hidden," neon, from the Greek word *neos*, meaning "new," and xenon, from the Greek word *xenos*, meaning "strange." The reason for these unusual names is because these were unexpected discoveries.

K ◆ William Ramsay was awarded the 1904 Nobel Prize in Chemistry for his discovery of a series of noble gases, including krypton.

◆ Krypton's concentration in Earth's atmosphere is about 1 part per million (ppm).

◆ Krypton is used in many different types of light bulbs, including:

    ✎ Neon light bulbs, as it allows for the creation of different colors.
    ✎ Fluorescent light bulbs, as it increases the energy efficiency.
    ✎ Bright incandescent light bulbs, as it helps protect the hot tungsten filament from burning out too quickly, particularly in products like projectors.

꙳ Krypton fluoride excimer lasers are used for semiconductor manufacturing and laser surgery because of their intense ultraviolet (UV) light beam. The wavelength of UV light allows for exceptionally high precision in making cuts and incisions.

꙳ And, of course, Krypton is the birthplace of (the fictional social activist comic-book hero) Superman, whose Kryptonian name was *Kal-El*.

K-T Extinction

❧ The K-T extinction refers to the worldwide event that took place 65 million years ago when the dinosaurs—as well as almost all of the large vertebrates on Earth, land, sea, and air—became extinct.

❧ The "K" stands for "Cretaceous." You read it right: K, not a C, from the German word *kreidezeit*, meaning "chalk time." The "T" stands for "tertiary."

❧ The reason for this mass extinction, and presumably sudden and violent end to the Cretaceous period, is still hotly contested in the

TIME OUT, REX. WE MAY HAVE TO RETHINK OUR SURVIVAL STRATEGIES.

THE *K-T* EXTINCTION

scientific community. The two hypotheses that have substantive evidence include a large asteroid impact and massive volcanic eruptions.

🐦 The asteroid impact theory is strongly buoyed by the discovery of what is believed to be the impact crater. About 111 miles (180 km) across, the crater is located within the Yucatán peninsula of Mexico. The rock is about 65 million years old, so the timing is perfect.

🐦 The strong evidence for the gigantic volcanic eruptions theory is compelling as well. The timing is right, as the Deccan eruptions began suddenly just before the K-T boundary, about 65 million years ago. Plus, the estimates of the massive fire fountains generated suggest that aerosols and ash would easily have been carried into the stratosphere, thereby blocking out the sun, which could easily have resulted in, or at least contributed to, a massive extinction of plant, animal, and sea life.

🐦 Additionally, according to a study just published in 2012, the famous Tyrannosaurus Rex FMNH PR2081 (better known as "Sue") is now believed to have died as the result of being bitten by a bird infected with trichomonosis—a type of parasite.

🐦 So what caused the K-T extinction? My personal belief, and the belief of many scientists, is that a confluence of events caused the extinction, as opposed to a single smoking gun. These events include the Yucatán peninsula asteroid, the Deccan volcanoes, and probably additional factors, including various insidious pathogens like viruses, bacteria, and parasites that have yet to be scientifically verified as widespread and/or discovered.

Lake Baikal

 Lake Baikal is to Russia what the Grand Canyon is to the United States: A magnificent natural resource that instills national pride and awe.

 Located in Eastern Siberia, Lake Baikal is over 25 million years old, making it the oldest lake in the entire world!

 A staggering 395 miles (636 km) long, and nearly 50 miles wide (79 km), Lake Baikal is the largest freshwater lake in the world.

 Lake Baikal is the world's deepest lake, at over 5,300 feet (1,600 m).

 Lake Baikal contains as much fresh water as the five Great Lakes of North America combined.

 How much water are we talking? Six quadrillion gallons (23,615.39 km³)!

 In other words, Lake Baikal contains roughly one-fifth of the total surface freshwater in the world!

 Despite the lake's great depth, its water is incredibly well oxygenated throughout, creating unique biological habitats.

 In fact, 80% of the species living in Lake Baikal are found nowhere else on Earth—including the world's only freshwater seal.

 For five months of the year, an ice sheet covers Lake Baikal and is over 3 ft (1 m) thick. This *is* Siberia!

Landslides

🐦 Visually, a landslide resembles a snow avalanche, only with a louder rumbling noise, and is capable of generating enough force and momentum to wipe out anything in its path.

🐦 The world's biggest *prehistoric* landslide (discovered so far) is the Saidmarreh landslide, located in southwestern Iran. It's believed that upward of 50 billion tons of rock moved in that single event!

🐦 The world's biggest *historic* landslide occurred during the 1980 eruption of Mount St. Helens, a volcano in the Cascade Mountain Range.

🐦 Landslides are often more damaging and deadly than the triggering event (e.g., the 1964 Alaska earthquake-induced landslides and the 1980 Mount St. Helens volcanic debris flow).

🐦 The Mount St. Helens' rock slide-debris avalanche was big enough to fill 250 million dump trucks, traveled 14 miles (22 km), killed 57 people, destroyed nine highway bridges, numerous buildings, and miles of highways, roads, and railroads—and caused an estimated 1.1 billion dollars in damage.

🐦 While often accompanying earthquakes, floods, storm surges, hurricanes, wildfires, or volcanic activity, landslides can occur on any terrain given the right conditions of soil, moisture, and the angle of slope.

🐦 Landslides can also be triggered by human-made causes, such as grading, terrain cutting and filling, deforestation, blasting, vibrations from machinery and traffic, and excessive development.

L

Latitude

❧ The tilt of Earth varies between about 22.1 and 24.5° in a 41,000-year cycle. Thus, the polar circles of latitude (like the Arctic and Antarctic) and the tropical circles of latitude (like the Tropics of Cancer and Capricorn) actually *move* every year by about 49 feet (15 m)!

❧ A "circle of latitude" is an imaginary east-west circle. These circles are often called "parallels" because, yes, you guessed it: they run parallel to each other.

❧ There are five named circles of latitude from north to south: The Arctic Circle, the Tropic of Cancer, the equator, the Tropic of Capricorn, and the Antarctic Circle.

❧ The positions of the polar and tropical circles of latitude, excluding the equator, depend on the tilt of Earth's axis relative to the plane of its orbit around the Sun.

❧ The Tropic of Cancer is defined as the northernmost latitude at which the Sun can appear directly overhead. Similarly, the Tropic of Capricorn is defined as the southernmost latitude at which the Sun can appear directly overhead. These are currently about 23° of latitude north or south of the equator. Remember, they're MOVING!

❧ The Arctic Circle, currently at about latitude 66° and 33' North, is defined as the start of the area where the Sun doesn't completely set on the June solstice or rise on the December solstice.

The Antarctic Circle, currently at latitude 66° and 33' South, is defined as the start of the area where the Sun doesn't completely rise on the June solstice or set on the December solstice.

There are venerable maritime traditions associated with crossing the various fixed lines of latitude by ship. The most commonly known "crossing the line" ceremony is admission to the Court of King Neptune and bestowal of the title "shellback" to sailors crossing the equator. A title of "golden shellback" is a special recognition for those crossing the equator at the International Date Line, and an "emerald shellback" is the title for those for crossing the equator at the Greenwich Prime Meridian!

Sailing across the Arctic Circle makes one a member of "The Order of the Blue Nose," and crossing the Antarctic Circle grants a sailor entry to "The Order of the Red Nose." (… Hmmm, I thought Rudolph the red-nosed reindeer lives in the North Pole! Maybe not!)

Lead

  ❧ With a history dating back at least 7,000 years, lead is thought to be the oldest metal.

  ❧ From the Latin word *plumbum*, meaning "waterworks," the chemical symbol for lead is Pb.

  ❧ The ancient Romans used lead to make water pipes. Since lead is a cumulative poison, the fall of the Roman Empire has been blamed, in part, on lead in the water supply.

  ❧ "Plumbism" is the medical term for lead poisoning.

  ❧ Lead exposure in children is associated with kidney damage, reduced IQ, slowed body growth, academic failure, as well as behavior, attention, and hearing problems.

  ❧ In the United States, lead-based paints were banned for use in housing in 1978.

  ❧ Leaded gasoline was phased out from 1993 and banned in 1995. However, fuel containing lead may continue to be sold for off-road uses, including aircraft, racing cars, farm equipment, and marine engines.

  ❧ According the National Institutes of Health, "Lead poisoning has been linked to lower IQ scores in children exposed to even low levels of lead."

꩜ Blood lead levels (BLL) in children over 10 micrograms per deciliter (10 mg/dL) are reason for concern. (A microgram is one millionth of a gram; a deciliter is about half a cup of liquid.)

꩜ As of May 2012 (and for the first time in 20 years!), the CDC has lowered the level at which a child is deemed to have lead poisoning from 10 micrograms per deciliter to 5 micrograms of lead per deciliter of blood.

꩜ The CDC also dropped the term "level of concern" because there is no exposure below which they have no concern and will use the new value to drive primary prevention efforts.

꩜ In other words, *any* lead exposure is cause for concern.

꩜ In adults, lead exposure can increase blood pressure, cause fertility problems, nerve disorders, muscle and joint pain, irritability, and memory or concentration problems.

꩜ Lead is still used for a vast range of purposes. It provides X-ray shielding, serves as fishing weights, and is still used in batteries.

Lichtenberg,
Georg Christoph

❧ Georg Christoph Lichtenberg (July 1, 1742 to February 24, 1799) is one of the coolest historical figures. He's a physicist, professor, writer—satire was his specialty—and Anglophile, and you've probably never heard of him!

❧ Ever heard of Lichtenberg Figures? Well, you've certainly seen them ... but perhaps just didn't know what they were called. They're those uber-awesome branching electric discharges that look like magical, electrical trees that are sometimes preserved on the surface or interior of a solid dielectric. Lichtenberg discovered and studied them! And today, they're named in his honor.

L

❧ In 1777, Lichtenberg discovered the basic principle upon which modern-day photocopying—called xerography—was based (i.e., dry electrostatic printing). He discovered this principle as an unexpected finding associated with his building a capacitive generator (called an electrophorus), to generate high-voltage static electricity via the process of induction.

❧ It was Lichtenberg who initiated the concept of a standardized paper size system, known as ISO 216, which is still used today across the globe, sans Canada and the United States.

❧ Prior to Lichtenberg, scientists would share their findings via oral lectures (yawn!). Lichtenberg was one of the first—and some sources say the very first—scientists to make lectures actual demonstrations and engage his audiences using visual aids and other apparatus! Such noteworthy attendees to his popular lectures included Carl Friedrich Gauss, one of the most influential mathematicians of all time, and Alessandro Volta, who invented the battery. Yup, pretty impressive.

❧ Lichtenberg's writing—which he called "waste books"—influenced such notable figures as Nietzsche, Freud, Goethe, Kant, Schopenhauer, and Qian Zhongshu—and they refer to him in their own works! Can you imagine?

❧ Lichtenberg's waste books addressed a bevy of topics and musings ranging from his thoughts on human nature, quotations that struck his fancy, and titles of books to read, to brilliant insights concerning the scientific method of inquiry. He started writing his waste books as a student in 1765 and continued writing them up until his death in 1799.

❧ An early adopter of technology, Lichtenberg was one of the first people in Germany to install Benjamin Franklin's lightning rod in his own home.

❧ Lichtenberg was the very first scientist to hold a professorship explicitly dedicated to experimental physics.

❧ There's also a crater on the Moon named in his honor—the Lichtenberg Crater—located north of the Briggs Crater, in the western part of the Oceanus Procellarum.

Lightning

~❧ A lightning bolt takes only a few thousandths of a second to split through the air.

~❧ Because electricity follows the shortest route, most lightning bolts are close to vertical.

L

---✦✦✧---

✦ As lightning connects to the ground from the clouds, a second lightning bolt will return from the ground to the clouds (the "bounce-back bolt"), following the same channel as the first strike.

✦ The heat from the electricity of the ground-to-cloud second lightning strike (or "bounce-back bolt") raises the temperature of the surrounding air to about 48,632°F (27,000°C)!

✦ Since the lightning takes so little time to go from point A to point B, the heated air has no time to expand.

✦ The heated air is compressed, raising the air from 10 to 100 times the normal atmospheric pressure.

✦ The compressed air explodes outward from the channel, forming a shock wave of compressed particles in every direction.

✦ Like an explosion, the rapidly expanding waves of compressed air create a loud, booming burst of noise (i.e., thunder).

✦ Lightning doesn't always create thunder (e.g., in April 1885, five lightning bolts struck the Washington Monument during a thunderstorm, yet no thunder was heard).

✦ To estimate approximately how close lightning may be to you, count the seconds between the flash and the thunderclap. Each second represents about 984.25 feet/300 m. [*See* Thunder, *page 228.*]

Liver

❧ Affecting nearly every physiological process in the human body, and weighing in at an impressive 3 to 4 pounds (1.4 to 1.8 kg), the human liver performs over 500 functions!

❧ The average healthy adult's liver weighs about the same as, and sometimes up to a pound *more* than, a female or male human brain, which weigh about 2.8 to 3 pounds (about 1.28 kg) respectively. However, size doesn't necessarily matter: Albert Einstein's brain weighed only 2.71 pounds (1.23 kg).

❧ About the size of a football, the liver is the largest organ *inside* the human body (the skin is the largest organ, overall). Even if the liver is damaged upward of 90%, the liver is one of the only human organs that can regenerate.

❧ The liver is comprised of about 96% water! Yet, it can convert, store, and release sugars (energy) into your body as needed, thereby playing a key role in blood sugar regulation. It can also produce bile, which breaks down fats in the foods you eat and destroy dangerous toxins, bacteria, and other microbes that can potentially pose a health risk or be fatal.

❧ To safely filter out, break down, and eliminate the constant onslaught of harmful toxins flowing through your body—which includes environmental toxins such as car exhaust and pesticides—the liver processes over a quart (about a liter) of blood every single minute!

L

Longitude

~✤ The International Date Line does not exactly follow the 180° longitude line: Rather, it zigs and zags between Russia and Alaska and around certain island groups in the Pacific Ocean.

~✤ The International Meridian Conference of 1884 established the Prime Meridian (0° longitude) passing through the Royal Observatory, Greenwich, in the United Kingdom.

L

꙳ A "meridian" is simply an arbitrary north-south line selected as the "zero" reference line for astronomical observations or cartographic (map-making) measurements.

꙳ Just as the equator divides the Northern and Southern hemispheres, the Prime Meridian and its counterpart on the other side of the world (180° longitude) divide the Eastern and Western hemispheres of Earth.

꙳ The Prime Meridian at Greenwich was defined with a special telescope called a "transit circle," built in 1850 by Sir George Biddell Airy, the seventh Astronomer Royal.

꙳ In fact, the intersection of the crosshairs viewed in the eyepiece of the transit circle was used to define 0° longitude in 1884.

꙳ If you stand on the Prime Meridian line in Greenwich and check your GPS location, it will NOT read 0° longitude.

꙳ The Global Positioning System (GPS)—which we count upon to navigate our lives—actually uses a slightly different coordinate system of references, called the WGS 84. What's the difference? To read zero longitude on your GPS, you'd have to stand about 336.3 feet (102.5 m) to the east of the Greenwich Prime Meridian.

Mars

~❧ Called the "red planet" because of its high surface iron oxide content, which, of course, is reddish, Mars is the fourth planet from the Sun and is 4,217 miles (6,785 km) in diameter—about half the diameter of Earth.

~❧ Just like Earth, Mars has a midnight sun—meaning, it remains visible for a continuous 24 hours—at the poles during the summer. Mars also shares the opposite phenomenon in the winter, called "polar night," where the Sun stays below the horizon throughout the day.

~❧ Mars is about 128 million miles (205 million km) from the Sun. Its revolution period—how fast it revolves around the Sun—is 1.88 years, or about 22½ months.

M

❧ The length of a single day on Mars is 24 hours and 37 minutes long, which is similar to the length of a day here on Earth.

❧ Mars has higher mountains and deeper canyons than *any other planet in our solar system*. It also has the biggest volcano, named Olympus Mons, which is almost three times the size of Mount Everest (the tallest mountain on Earth).

❧ Named for the mythical horses that drew Mars' chariot, the planet's two moons—which aren't particularly round at all—are named Deimos and Phobos. "Deimos" means "panic," and "Phobos" means "fear."

❧ Compared to our atmosphere, Mars has a rather thin atmosphere that is comprised mostly of carbon dioxide (95.3%), nitrogen (2.7%), and argon (1.6%). And while colonization is a topic of serious discussion in some scientific circles, oxygen makes up only 0.13% of Mars' atmosphere.

❧ What's in a name? Because the color of Mars looks like blood, the planet was named for the Roman god of war. Incidentally, Mars was also Venus' lover, the father of Romulus and Remus, and the son of Juno and Jupiter. The month of March was named after him.

Marsupials

✦ All infant marsupials are called "joeys," not just kangaroos.

✦ Although today marsupials are most widely represented in Australia, genetic analysis indicates that all living Australian marsupials evolved in South America from a common ancestor, which later migrated to Australia.

✦ Neither marsupials nor monotremes have the bundle of nervous tissue that links the left and right sides of the brain—called the corpus callosum—that most mammals possess, although it's thought that some other brain structures, such as the anterior commissure, may serve this function.

✦ Koalas are one of the few mammals (other than primates) that have fingerprints!

✦ A now-extinct giant wombat-shaped marsupial called the *Diprotodon optatum* grew up to 10 feet (3 m) long, stood 6½ feet (2 m) at the shoulder, and weighed in at about 4,500 pounds (2,000 kg)!

M

Mendacity

🐦 From the Latin word *mendacitas*, meaning "lying," "mendacity" is a noun that means falsehood, prevarication. More simply put, "mendacity" is a lie.

🐦 Can you tell if a person is lying? Well, maybe—read on!

🐦 Eye pupil dilation is an involuntary, unconscious nervous system response to lying, which may expose a lie.

🐦 There are specific facial expressions that flash on a person's face for just a split second that can be revealing. These superfast expressions are called "micro expressions."

M 🐦 There are a number of facial expression-based scientific methodologies used for lie detection, the specifics of which are well-guarded secrets within certain "special operations sectors" for one obvious reason: They don't want people learn how to beat the system.

🐦 Contrary to popular belief, when a person's eyes shift right or left it's not a dead give-away that that person is lying. It's far more complicated than that. The right verses left brain distinction isn't nearly as clear cut as once believed; plus, if a person has told the lie often enough there may be no detectable stress response. Lying is also an effortless event for some people, requiring little or no forethought.

◈ Don't count on an increase or decrease in eye contact—such as when a person looks you dead in the eye or when they look away—as an indicator of veracity. Studies have found that this is not a clear indicator.

◈ That said, as a general rule, mendacious people do tend to make fewer speech errors than those telling the truth.

◈ Liars also tend to blink less frequently and take greater pauses while speaking ... but not always!

◈ Truth be told? Statements made by people prone to mendacity should be taken with a grain of salt.

◈ ... Except when that's not true.

Mercury

🐦 Mercury is the planet closest to the Sun. Covered with craters, Mercury is the smallest planet in our solar system. In fact, it's only slightly larger than Earth's moon.

🐦 Mercury has almost no atmosphere thanks to gravity, which is too weak to hold an atmosphere of any real substance. Mercury's solar winds also contribute to its lack of atmosphere, and the winds literally continuously blow away any chance of an atmosphere forming.

🐦 Without any atmosphere to speak of, Mercury has no weather.

🐦 Mercury is only 28.6 million miles (46.0 million km) from the Sun. Hence, Mercury's revolution period, or how fast it revolves around the Sun, is a scant 0.24 years, which means it speeds around the Sun in only about three months.

🐦 What's in a name? In mythology, Mercury is the messenger god—also called "Winged Mercury"—and he was also the god of games, business, and storytelling.

🐦 Mercury is sometimes referred to as a "morning star" because it shines very brightly just before the Sun rises. It's also sometimes called an "evening star" because it is often briefly visible just after the Sun sets.

M

Metal Alloys

 An alloy, by definition, is a blend of two or more metals.

 Bronze is an alloy of copper and tin.

 Brass is a combination of copper and zinc.

 Steel is iron which has carbon added to it.

 Some specialized steel alloys have other metals added to produce specific properties, such as stainless steel, which has chromium added to reduce corrosion, rust, and stains.

 Depending upon the color, the metal alloy gold may have copper to make pink gold, platinum for white gold, and a combination of silver cadmium (and sometimes a touch of copper) for green gold. Only 24-karat gold, which is 100% pure, has no other added metals.

M

 Electrum is a lesser-known metal that was referred to as "white gold" by the Greeks. It was used to make drinking vessels and coins as far back as 3000 BC and is a naturally occurring alloy of gold and silver with trace amounts of copper and other metals.

 Solder is typically a blend of lead and tin.

 Pewter is mostly tin with copper, antimony, bismuth, and lead.

Meteoroids, Meteors, and Meteorites

🖎 Comprised of rock and particles of solar system debris, meteoroids are objects out in space that we're unlikely to see unaided by a telescope … if at all. Meteoroids range in size from a speck of dust to a rock around 10 yards (about 10 m) in diameter,

🖎 The fastest meteoroids in our solar system travel at speeds of around 26 miles per second (about 42 km per second).

🖎 A meteor was once a meteoroid. Once it has entered our atmosphere, we see it as a bright flash of light streaking through the sky.

M

🖎 Commonly (or perhaps wistfully) called a "shooting star," the shooting light from a meteor appears as a result of it burning up as it passes through Earth's atmosphere.

🖎 A meteorite is what we call a meteor once it reaches the ground.

🖎 When a meteorite streaking through the sky falls and is then recovered, it's called a "fall." If the streaking and falling isn't witnessed, but a meteorite is nonetheless discovered, it's called a "find."

🖎 According to the most recent estimates, there have been approximately 1,000 collected "falls" and an impressive 40,000 "finds."

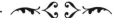

On rare occasions, a meteoroid enters Earth's atmosphere but makes it safely out: These are called "Earth-grazing fireballs."

A "meteor shower" is a term used to describe the process when many meteors from the same general part of the sky appear within a close time frame.

Meteorology is not the science of meteors!

Monotremes

🐾 Monotremes are mammals that lay eggs instead of giving birth to live young like marsupials do. Marsupials give birth to live young that continue to develop in the pouch.

🐾 Surviving examples of monotremes include the platypus and echidnas (spiny anteaters). The bill of a platypus contains an electrosensory organ that helps it find prey underwater as it swims with its eyes, ears, and nostrils closed.

🐾 While searching for prey underwater, platypuses swim with their eyes, ears, and nostrils closed! How then, do they successfully find their next meal? Their bills are covered with thousands of electric-detecting receptors (which can sense the electrical fields of potential prey) and mechano-receptors (which detect movement). Very cool!

🐾 Like reptiles and birds, monotremes have a single external opening for the reproductive, urinary, and digestive systems called the "cloaca."

🐾 According to aboriginal legend, the very first platypus was born after a young female duck mated with a lonely and persuasive water rat. The duck's offspring had their mother's bill and webbed feet and their father's four legs and handsome brown fur.

M

LEGENDARY ORIGIN OF THE PLATYPUS

Moon

~❧ A rocky body about 2,160 miles (3,476 km) in diameter, the Moon orbits Earth at an average distance of 237,612 miles (382,400 km). Why "average"? The Moon's distance from Earth varies because it follows an elliptical orbit instead of a circular one. It ranges from 221,457 miles/ 356,400 km at perigee (closest to Earth) and 238,606 miles/384,000 km at apogee (farthest).

~❧ The Moon's distance from Earth varies because it follows an elliptical orbit instead of a circular one. It ranges from 221,457 miles (356,400 km) at perigee—closest to Earth—and 238,606 miles (384,000 km) at apogee— farthest from Earth.

M ~❧ We enjoy a "super moon" when the Moon is at perigee, during which time the Moon appears about 14% bigger and 30% brighter.

~❧ The same side of the Moon always faces Earth. The Moon's rotation period is synchronous with its revolution period around Earth.

~❧ Hot by day and very cold by night, the average surface temperature of the Moon is 224°F (107°C) during the day and -243°F (-153°C) at night.

~❧ The dark-looking "seas" on the Moon are really smoother and darker-colored rock that came from volcanic flows early in the history of the Moon after most of the cratering had occurred. These dark spots are called "Maria."

❧ There are eight phases of the Moon, which include: new Moon, waxing crescent, first quarter moon, waxing gibbous, full moon, waning gibbous, third-quarter moon, and waning crescent.

❧ Just like Earth, half of the Moon is lit by the Sun while the other half is in darkness.

❧ Like the rest of the solar system, the moon is about 4½ billion years old. Most of the cratering occurred in the first billion years of its existence, when leftover debris from planet formation was abundant.

❧ We only see the Moon because sunlight reflects back to us from its surface.

❧ The Moon is very gradually moving away from Earth because of a loss of orbital energy to gravity from Earth. Long ago, the Moon regularly looked about three times larger in the sky because it was closer to Earth.

Moon Names

🐦 Ever wonder where the term "harvest moon" came from? The different names for the Moon were created by the Algonquian tribes of Native Americans—and each for a very specific reason!

M

- 🌙 **January:** Wolf Moon—Hungry wolf packs howled at night during this time.
- 🌙 **February:** Snow Moon—Heaviest snowfalls occurred in the middle of winter.
- 🌙 **March:** Worm Moon—This was the start of spring, as earthworms (and the robins that eat them!) began to appear.
- 🌙 **April:** Pink Moon—An early spring flower called "moss pink" started to bloom.
- 🌙 **May:** Flower Moon—Many types of flowers bloom in May.
- 🌙 **June:** Strawberry Moon—Strawberries were ready to be picked and eaten.
- 🌙 **July:** Buck Moon—New antlers of buck deer, coated with velvety fur, began to form.
- 🌙 **August:** Sturgeon Moon—Sturgeon, a large fish found in the Great Lakes, were easily caught at this time of year.
- 🌙 **September:** Harvest Moon—Farmers could continue harvesting until after sunset by the light of the harvest moon.
- 🌙 **October:** Hunter's Moon—Hunters tracked and killed prey by moonlight, stockpiling food for the coming winter.
- 🌙 **November:** Beaver Moon—Time to set beaver traps before the swamps froze, to make sure of a supply of warm winter furs.
- 🌙 **December:** Cold Moon—The cold of winter sets in.

Neptune

🐦 A brilliant and beautiful, rich, periwinkle-blue planet, and the smallest of the four gas giants, Neptune is the eighth and farthest official planet from the Sun. (Pluto was demoted to a dwarf planet.)

🐦 Discovered by Voyager II, Neptune has a huge white cloud in its atmosphere that seriously scoots around! Thus, its name is "Scooter." Really.

🐦 Not unlike Jupiter's Giant Red Spot, Neptune has a Great Dark Spot— and a Small Dark Spot! But unlike Jupiter, Neptune's Giant Red Spot appears and disappears.

🐦 Similar to the hole in Earth's ozone layer, the great dark spot of Neptune is also thought to be a hole in its methane cloud layer.

N

🐦 The winds of the great dark spot are believed to have reached 1,200 mph (approx. 1,931 km/h)—which would mean that Neptune has the strongest winds in our solar system.

🐦 And particularly cool, those incredible winds sweep the dark spots (big and small) "backward" around the planet. Yes, in the opposite direction of the planet Neptune's orbital spin!

🐦 Neptune has six rings circling the planet, which are believed to be fairly new, as compared to the other ringed planets.

◦ Neptune has 13 known moons; it's about 2.77 billion miles (4.46 billion km) from the Sun; and has a revolution period (i.e., how fast it revolves around the Sun) of 165 of our Earth years.

◦ What's in a name? Named for the gorgeous planet's blue color (like the ocean), Neptune was the Roman god of the waters (Greek Poseidon), and brother of Pluto (Hades), and Jupiter (Zeus).

Neurons

🍂 While "neuron" can be the name for nerve cell, people generally use to the term to refer to brain cells.

🍂 The average number of neurons in the human brain is 100 billion.

🍂 The average number of neurons in an octopus brain is 300 billion.

🍂 The rate of neuron growth during development of a fetus in the womb is 50,000 neurons per minute.

🍂 The diameter of a neuron is 4 to 100 microns.

🍂 The animal with the longest axon of a neuron (i.e., the long, slender nerve fibers where electrical impulses travel to communicate information) is the giraffe—whose primary afferent axon (extending from their neck to their toes) is a whopping 15 feet (about 4.5 m).

N

🍂 The velocity of a signal transmitted through a neuron is 1.2 to 250 mph.

🍂 Get this: If we lined up all of the neurons in your average human, end to end, it would span upwards of 600 miles (or about 1,000 km)! The math? That's easy: 100,000,000,000 neurons x 10 microns (average length of neuron) equals about 600 miles (1,000 km).

Nobel Prize

🕊 Established in the will of scientist, author, and entrepreneur Alfred Nobel in 1895, the Nobel Prize is the pinnacle of honor that recognizes people for outstanding achievements in physics, chemistry, medicine, literature, and peace ... Alfred Nobel also invented dynamite.

🕊 Over 40 scientists have received Nobel Prizes in chemistry, physiology, or medicine for their work on DNA (deoxyribonucleic acid). In fact, DNA is responsible for more Nobel Prizes than any other single molecule.

🕊 Marie Sklodowska Curie, also known as Madame Curie, was the first woman to be awarded a Nobel Prize and the first person ever honored with two Nobel Prizes. In 1903, she received a Nobel Prize in physics for her groundbreaking research on the radiation phenomena, and in 1911 she received a Nobel Prize in chemistry for discovering radium and polonium.

N

🕊 In fact, one of the primary reasons Alfred Nobel established the prizes was the result of a newspaper mistakenly thinking he had died (but it was actually Alfred's brother, Ludvig, who had died). The obituary characterized Alfred as *"Le marchand de la mort est mort"* (The merchant of death is dead) and went on to say that Alfred had become "rich by finding ways to kill more people faster than ever before, died yesterday."

⬥ He was purportedly so distraught that this was to be his legacy that he left the bulk of his sizable estate to the establishment of the Nobel Prizes, specifically to be used to endow "prizes to those who, during the preceding year, shall have conferred the greatest benefit to mankind." This financial endowment was a particularly big deal, as it was incredibly unusual at that time to donate large sums of money for scientific and charitable purposes!

⬥ The first Nobel Peace Prizes were awarded in 1901. The back of the commemorative medal that has been struck for this auspicious occasion depicts a tunnel blasted by dynamite and a detonator or blasting cap! The front of the medal depicts a portrait of Alfred Nobel, with the Latin inscription *Creavit et promovit*, which means "He created and promoted."

Northern Lights: History

- Both Galileo Galilei and Pierre Gassendi are credited for first using the term aurora borealis, meaning "northern dawn," since both were describing the same September 12, 1621, light display!

- The earliest account of the Northern Lights is from observations made by the official astronomers of King Nebuchadnezzar II, c. 568/567 BC, which were made on a Babylonian clay tablet.

- Gallo-Roman historian Gregory of Tours (c. 538 to 594) described the Northern Lights as "... so bright that you might have thought that day was about to dawn."

N

- Benjamin Franklin described the Northern Lights, too! He posited that due to a concentration of electrical charges in the polar regions, intensified by the snow and other moisture, this overcharging caused a release of electrical illumination into the air.

- In 1896, Kristian Olaf Birkeland proposed the theory that electrons from sunspots triggered auroras, a theory that was originally dismissed by fellow scientists as fringe theory (a theory that lacks scientific integrity).

- It wasn't until 1967 that a NASA space probe provided proof that Birkeland's aurora theory was correct.

- Today, Birkeland is considered the "father of modern auroral science."

Northern Lights: How It Happens

❧ Called a coronal mass ejection (CME) by scientists, the aurora begins on the surface of the Sun when solar activity ejects a cloud of gas.

❧ If a CME reaches Earth, which takes two to three days, it collides with Earth's magnetic field.

❧ Quick side comment (because it's too cool not to share, and it's germane): If we could actually see the shape of our planet's invisible magnetic field, it would make Earth look like a giant comet with a fantastically long magnetic tail stretching a million miles behind us in the opposite direction of the Sun!

❧ When a CME collides with the magnetic field, it causes very cool and complex changes to happen to the magnetic tail region. These changes generate currents of charged particles, which then flow along lines of magnetic force into the polar regions. The polar regions then get a super energy boost in Earth's upper atmosphere! When those super-charged particles collide with oxygen and nitrogen atoms, that's what ultimately triggers the dazzling aurora borealis light show!

❧ The light displays may take many forms, including rippling curtains, pulsating globs, traveling pulses, or steady glows.

N

❧ In addition to the various shapes being affected by magnetic forces, the amazing patterns and colors depend on:

    ❧ The types of ions or atoms being energized as they collide with the atmosphere.

    ❧ The altitude: Blue violet/reds occur below 60 miles (100 km); bright greens are strongest between 60 to 50 miles (100 to 240 km); and above 150 miles (240 km) breathtaking ruby reds appear!

❧ Auroras tend to be more frequent and spectacular during high solar sunspot activity, which cycles about every 11 years.

❧ And so that there's no confusion, the term "polar lights" (aurora polaris) refers to the lights found in both the Northern and Southern hemispheres. The aurora borealis refers to the Northern Lights, and aurora australis refers to the Southern Lights.

Northern Lights: Legend and Lore

🐦 Ancient Inuit called the aurora *aqsarniit* (meaning "football player"), as they believed the souls of the dead were at play, using a walrus skull as a ball.

🐦 An Algonquin creation myth explains that when Nanahbozho, creator of Earth, finished his grand task of creating Earth, he chose to reside in the North. As a visual reminder to his people that he still thought of them, he built splendid fires, of which the Northern Lights are the reflections.

🐦 Ancient Chinese and European folklore described auroras as great dragons or serpents in the skies.

🐦 Scandinavia, Iceland, and Greenland lore often ascribe the aurora as *Bifröst*, a burning rainbow bridge by which the gods traveled between Heaven and Earth.

🐦 There's an Inuit legend that claims the Northern Lights are torches lit by spirits of the dead already residing in heaven who are guiding the feet of the new arrivals.

🐦 Similarly, there are Native American tribes who attributed the aurora to spirits carrying lanterns as they sought the souls of dead hunters.

🐦 Inuit legend of the lower Yukon River describes the aurora as the dance of animal spirits, especially those of deer, seals, salmon, and beluga.

N

❧ The Salteaus Indians of eastern Canada and the Kwakiutl and Tlingit of southeastern Alaska saw the Northern Lights as the spirits of people dancing.

❧ There are also many cultures that believe that a crackling, swishing, or hissing sound accompanies the aurora and that this noise is the voices of spirits trying to communicate with the people of Earth.

Nuclear Energy

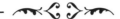

✤ Nuclear energy is energy in the nucleus, the core of an atom.

✤ Atoms are tiny particles that make up every object in the universe. There is enormous energy within the bonds that hold the nucleus together. Breaking those bonds releases that energy.

✤ The fuel most widely used by nuclear plants for nuclear fission is uranium, a non-renewable resource. [*See* Nuclear Fission, *page 162.*]

✤ A nuclear reactor is essentially a furnace in which energy is generated by a controlled fission chain reaction.

✤ The specific uranium isotope used in nuclear power plants (... and bombs) is called U-235. This *isotope*—which simply means "type" or "form"—has atoms that split more easily than other forms, such as U-238, which is the most stable isotope of uranium.

✤ A common metal found in rocks, most of the uranium used in the United States is mined in the western part of the country.

✤ The average nuclear power plant in the United States generates about 12.4 billion kilowatt-hours (kWh) annually.

✤ As of 2011, the average age of U.S. commercial nuclear reactors is about 31 years. The oldest operating reactors are Oyster Creek in New Jersey, and Nine Mile Point 1 in New York. Both entered commercial service on December 1, 1969.

N

Nuclear Fission

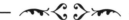

 The splitting of a uranium nucleus, which is massive, into two less massive fragments is called "nuclear fission." In a nutshell, here's how it works:

 A small particle called a neutron hits the uranium atom and splits it, releasing a great amount of energy as heat and radiation.

 Then, more neutrons are also released.

 These neutrons go on to bombard other uranium atoms, and the process repeats itself over and over again. This is called a chain reaction.

 That reaction releases huge amounts of energy.

 That energy can boil water to create steam, which, in turn, causes turbines to spin, generating electricity in a power plant.

N

Nuclear Fusion

🐦 Nuclear fusion can be described as the opposite of nuclear fission.

🐦 Nuclear fusion occurs when two nuclei with very low mass (and minimal binding energy) are fused to make a single, far more massive nucleus.

🐦 The term "thermonuclear" is used to describe the fusion reaction because extraordinarily high temperatures are necessary for this process to proceed.

🐦 To date, no nuclear fusion plants have ever been built because we've been unable to sustain the fusion reaction for commercial application from an economic standpoint.

N

🐦 The H-bomb, or the hydrogen bomb, uses an atomic fission explosion to trigger a nuclear fusion reaction!

🐦 The Sun and all active stars are fueled by the fusion reaction.

Nuts

(And Not So Nuts)

- All nuts are fruits.

- Almonds, Brazil nuts, cashews, and pistachios are actually fruit seeds, not nuts.

- Macadamia nuts and walnuts are really kernels.

- A peanut isn't a nut, either! It is both a legume and a seed.

It takes about 540 peanuts to make a 12-ounce (340-g) jar of peanut butter.

- Two past presidents of the United States have also been peanut farmers: Thomas Jefferson, who grew them at his home Monticello among other crops, and Jimmy Carter.

- The Queen of Sheba is said to have decreed the pistachio nut as the exclusive domain of royal consumption, even forbidding "commoners" from growing the nut for personal use.

- Historians generally agree that almonds, as well as dates, were among the earliest cultivated foods. They were mentioned in writings as far back as the *Five Books of Moses*.

Oceans:
Blue, Green, and Brown

🐦 There are several theories that explain why the ocean usually appears blue, and why it also appears as other colors, too:

🐦 Blue wavelengths are transmitted to greater depths of the ocean and are absorbed the least—compared to other colors—by the deep ocean water.

🐦 Water molecules scatter blue wavelengths by absorbing the light waves and then emanating the light waves in different directions. That's why what's mostly reflected back to our eyes are the blue wavelengths! Plus, particles in the water may help reflect blue light.

🐦 The ocean may also be reflecting the blue sky. However, this is prominent only at relatively low angles and when the water is smooth.

🐦 That said, the ocean can be many other colors depending on particles in the water, water depth, and skylight. Why? The colors we see depend on the reflection of the visible wavelengths of light to our eyes, and wavelengths of light pass through matter differently depending on the material's composition. For example, when the ocean appears green, this may be due to abundant plant life or sediment from rivers that flows into the ocean. The yellow pigments from plants mix with the blue light waves to make green.

🐦 Sometimes, parts of the oceans will look milky brown after a storm passes. This is because winds and currents associated with the storm churn up sand and sediment from the rivers that lead into the oceans.

Octopus

꘎ "Octopuses" and "octopi" are both acceptable plural forms for "octopus," according to *Merriam-Webster's Dictionary*.

꘎ Octopuses have three hearts, which are all located in their head.

꘎ With a brain configuration unlike any other species on the planet, octopuses have nine brains that include a central hub with upward of 180 million neurons, or brain cells, that are linked to eight additional neuron groupings at the base of each tentacle.

꘎ According to the Hawaiian creation myth, the octopus is the lone survivor of a previous version of our universe.

O

꘎ Octopuses are the only invertebrate known to use their well-developed vision to learn through observation. In scientific studies in a laboratory, scientists have demonstrated again and again that a freshly caught octopus can learn to perform an otherwise foreign task—like twisting a jar open—simply by watching another octopus that has learned to do the task.

꘎ Now understood by scientists to be emotional animals, octopuses turn red when angry and pale when frightened.

꘎ While varying by species, a female octopus lays between 20,000 and 100,000 rice-shaped eggs that she carefully guards, cleans, and aerates until they hatch, which may take two to 11 months. She dies shortly after her babies hatch.

❧ Each of an octopus's eight tentacles has 240 suction cups, making for a total of 1,920 suction cups to grasp its prey.

❧ If an octopus loses one of its arms, it grows back.

❧ Mostly nocturnal hunters, octopuses use their highly sensitive suckers—which contain thousands of chemical receptors—to touch and taste, allowing them to find food in small, dark crevices.

❧ Octopuses have sharp, beaklike mouths that make removing and eating shellfish and mollusks a breeze.

❧ The giant Pacific octopus (*Enteroctopus dofleini*) is the largest octopus species in the world. It can grow up to 30 feet (9 m) and has an arm span of over 14 feet (5.5 m).

❧ The smallest octopus—called the "Californian *Octopus micropyrsus*"—never gets more than 1 inch (2.5 cm) long and is known as the smallest octopus, although the star-sucker pygmy octopus (*Octopus wolfi*) is not far behind for the title.

❧ An octopus's skin has the ability to transform colors, shape, and texture according to need. The skin can exactly mimic the octopus's surroundings, as well as don the shapes of other sea creatures such as the deadly scorpion fish and a venomous sea snake.

Oenology

🍂 Oenology (pronounced *EE-nawl-oh-gee*) is the study and science of wine making. Viticulture is the study of grapevines and grape growing.

🍂 The time period when the grapes are harvested and delivered to the wineries for processing to be made into wine is called "crush," a process that involves pressing and crushing. Crush typically lasts 60 to 90 days.

🍂 Wine grapes are harvested when their sugar content is between 22–26%. White wines are typically on the lower end of the sugar content spectrum, and reds on the higher end. [*See* Wines: Red and White, *page 268.*]

🍂 All grapes are either harvested by hand or by machine. This process occurs in the fall, regardless of geographic location.

🍂 A ton of grapes typically yields 180 gallons, which will fill three standard 60-gallon (227-liter) wine barrels.

🍂 Generally speaking, a traditional barrel holds 60 gallons (227 liters).

🍂 Each barrel yields approximately 20 to 25 cases of wine, which translates to 240 to 300 (750 ml) bottles of wine.

🍂 A case of wine is 9 liters. So, if the bottles are the standard 750 ml, that's 12 bottles of wine per case. If the bottles are magnums, which are 1500 ml, the wine will yield six bottles of wine per case.

Olfactory System

The word "olfactory"—which means "related to the sense of smell"—comes from the Latin word *olfactorium*, meaning "nosegay," which is a flower bouquet.

The human nose can detect and distinguish between 4,000 and 10,000 or more molecules of inhaled smells.

Located in the front third of the nasal septum (the bone that divides your nostrils) is Jacobson's organ, which can detect trace quantities of chemicals and recognize subtle human-to-human communications emitted via pheromones. [*See* Jacobson's Organ, *page 198*.]

The word "olfaction" is a scientific term for sense of smell.

The olfactory system is closely tied in with the gustatory system, which is the sense of taste. This explains why food doesn't taste good when your nose is stuffed.

Children are born with a heightened sense of smell and taste that diminishes as they get older, which may help explain some children's tendency to be picky with food!

Thoughts, memories, and emotions can be triggered by smells primarily due to the olfactory system's interconnectedness with the limbic system in our brains.

➤ When you inhale chemical molecules, or smells, they dissolve in the mucous lining residing atop the olfactory epithelium. The olfactory epithelim has upward of 40 million nerve impulse-generating little hairlike receptors, called cilia—which then transmit the information to the olfactory bulb. Your olfactory bulb then transmits the information to both your limbic system and your neocortex—which are your "primitive" brain and your "higher mental functioning" brain, respectively.

Ophidiophobes' Nightmare

🐍 In 2009, deep down in a Columbian coalmine, a team of paleontologists discovered the remains of the largest snake ever found in the world: the *Titanoboa cerrejonensis*.

🐍 Scientists estimate that *Titanoboa* lived 58 to 60 million years ago, which means it lived after the K-T extinction—the mass extinction that took place 65 million years ago that wiped out the dinosaurs, as well as almost all of the large vertebrates on Earth and land, in the sea and air. [*See* K-T Extinction, *page 122.*]

🐍 So, how big are we talking? Measuring 48 feet (14.6 m) and weighing in at a jaw-dropping 2,500 pounds (1,134 kg), this snake is longer than T. rex and easily dwarfs most living snakes today!

🐍 *Titanoboa* was so thick in girth it would have had trouble squeezing through your average doorway.

🐍 The snake that previously held the "largest snake" title was the *Gigantophis* of the Eocene period, about 40 million years ago, that measured over 33 feet (10 m).

🐍 The reticulated python—currently the world's largest living snake and the world's longest living reptile—stretches just barely more than half the length of the *Titanoboa*!

O

🐌 The green anaconda—the world's heaviest known species of snake to date—isn't even a paltry one-tenth the weight of *Titanoboa*. The heaviest scientifically documented green anaconda weighed *just* 215 pounds (97.5 kg).

🐌 Since snakes are cold-blooded reptiles, and the bigger the snake, the more heat energy they need to survive, the fossils of ancient snakes are like paleothermometers to scientists, providing amazing clues as to the possible temperature ranges of prehistoric times.

🐌 The insanely gigantic size of *Titanoboa*, for example, suggests to scientists that the average temperature during that time was likely higher than previously thought.

🐌 So, what's an ophidiophobe? A person who is scared to death of snakes!

Orange Roughy

🐟 When alive and well, orange roughy—also called "sea perch"—are actually a bright red. They pale to orange after they die.

🐟 Orange roughy are one of the longest-living marine species, averaging a lifespan of 120 to 130 years, although some have lived upward of 150 years. So, if you eat orange roughy, it might be from a fish older than your grandparents!

🐟 A deep-sea fish, the orange roughy is the largest of the slime head family (*Trachichthyidae*) and averages about 30 inches (75 cm) in length. It can weigh upward of 15 pounds (7 kg).

🐟 Orange roughy have a low reproductive rate (also referred to as "fecundity"), reproduce late in life—they don't reach sexual maturity until they're 23 to 32 years old—spawn irregularly, and have a very low egg count of 40,000 to 60,000. Thus, they are particularly vulnerable to overfishing.

O

🐟 Due to their high levels of mercury, the environmental defense fund has issued a health advisory concerning the consumption of orange roughy.

🐟 According to the Australian government's Department of Sustainability, Environment, Water, Population, and Communities: "Orange roughy's biological characteristics result in the species having very low resilience to fishing as the likelihood of being caught prior to reproduction is much higher in comparison to other fish species … Commercial trawling has been identified as a key threat to orange roughy populations."

Organic

❧ Something is considered organic if it contains carbon molecules, which includes pretty much every living thing (and previously living thing) on the planet.

❧ Every organic *compound* contains carbon, too. Aside from water, every living organism on the planet is primarily comprised of organic compounds.

❧ Organic chemistry is the study of chemical reactions involving molecules that contain carbon.

❧ There are also non-living things that are technically organic, like diamonds (those that are detritus and eclogitic), and coal and petroleum, which are plant and animal decay.

❧ The state of California requires any product sold and marketed as organic to contain a minimum of 70% organic content.

❧ In the United States, products must meet the standards of the USDA's National Organic Program (NOP) to be considered organic.

❧ The NOP is responsible for the regulatory framework that classifies products as organic. It regulates organic labeling and classification through administration and enforcement like the Organic Foods Production Act (OFPA) of 1990 and regulations in Title 7, Part 205 of the Code of Federal Regulation.

Owls

- Owls belong to the order *Strigiformes*, and there are over 200 species of owls.

- Owls have an 80% success rate for catching their prey.

- The structure of an owl's foot is usually zygodactyl, meaning two toes face forward and two toes face backward. Plus, the outer toe is capable of pivoting fore and aft (from the two/two configuration to three toes in front and one in back), which helps the owl not only capture his prey, but also hold onto it.

- An owl's amazing sense of hearing is aided by a flattened facial disk that optimizes funneling sounds to his ears, much like how a satellite dish functions.

- Contrary to popular belief, an owl can't turn its head completely backward. It can turn its head 135 degrees in either direction, thanks to its 14 neck bones and a swiveling bone structure at the base of its neck. Most mammals, including people, have only seven neck bones. Even giraffes have only seven neck bones!

- While they can't turn their head backward, they do enjoy a 270-degree field of view. While owls usually have outstanding night vision, they're also generally farsighted, so they have great difficulty seeing anything that's within close range.

O

To compensate for this shortcoming, owls have filoplumes, which are hairlike feathers, on their beak that function as combination pressure, vibration, and tactile monitoring receptors.

Owls don't have teeth. They must either swallow small prey whole or tear larger prey into smaller pieces before swallowing. Later, they regurgitate "pellets" that contain indigestible material such as bone, fur, and feathers.

Most owls are nocturnal, but not all. The pygmy owl is active either in the early morning or at dusk, and short-eared owls are active during the day.

Pacific Ocean

🐚 The Pacific Ocean covers nearly one-third of Earth's surface from Asia to the Americas, with a total volume of approximately 149 million cubic miles (622 million km³).

🐚 How big is that? Well, the Pacific Ocean would fill over 250 *trillion* Olympic-size swimming pools!

🐚 Spanning approximately 11,000 miles (17,700 km) from east to west at the equator, there is no ocean deeper or vaster than the Pacific.

🐚 The Pacific Ocean is more than twice the width of the Atlantic Ocean.

🐚 If you merged together all of the continents and landmasses on the planet, the Pacific Ocean would not only still cover more surface area than the combined world's land, there'd be enough room to add a second continent of Africa!

🐚 There are over 25,000 islands located in the Pacific Ocean.

🐚 Because of its expansive size, the islands of the Pacific are the most isolated in the world, with some more than 4,000 miles (6,400 km) away from the nearest continent. Pitcairn Island, for example, lies smack in the center of the vast ocean, and the better-known Easter Island's closest continent is about 2,000 miles (3,200 km) away.

P

❧ Generally speaking, the Pacific Ocean in the Northern Hemisphere moves clockwise, and in the Southern Hemisphere moves counterclockwise.

❧ It was Ferdinand Magellan who, in 1521, gave the Pacific Ocean its name (well, the name that we use today), which in his native tongue of Portuguese means "peaceful sea."

❧ In 1513, the Spanish explorer Vasco Núñez de Balboa had named the Pacific Ocean *"Mar del Sur,"* meaning "South Sea."

Pathogens

❧ From the Greek word *pathos*, meaning "suffering, passion," and *genēs*, meaning "producer of," a pathogen is defined as any infectious disease-producing agent like viruses, bacteria, parasites, fungus, and prions.

❧ For nearly every type of living cell—including the cells of bacteria, fungi, plants, animals, and people—there is a virus that can infect it.

❧ There are some viruses we can prevent through vaccination, such as measles, polio, and certain strains of the human papillomavirus virus (HPV), the most common cause of cervical cancer.

❧ We have never—*not ever*—successfully "cured" any virus, but we're working on it.

P

❧ The Special Pathogens Branch of the CDC (Centers for Disease Control and Prevention) is the division dedicated to working with Biosafety Level 4 (BSL-4) viruses.

❧ Biosafety Level 4 viruses are so highly pathogenic, such as Ebola hemorrhagic fever, Lassa fever, and Hantavirus pulmonary syndrome (HPS), that they require super special handling in even more super specialized laboratory facilities that are painstakingly designed to safely contain them (G-d willing.)

🐦 There are countless examples of good bacteria, but pathogenic bacteria can make us very sick, can be highly toxic, and even deadly. Examples of pathogenic bacteria include the Bubonic plague, Escherichia coli (E. coli), salmonella, clostridium botulinum (botulism), cholera, diphtheria, meningococci (meningitis), syphilis, tetanus, tuberculosis, and typhoid fever.

🐦 Parasites are organisms that live off of a host to the host's detriment. Examples include ticks, which carry Lyme's disease; worms and flukes like the tapeworm, which can cause schistosomiasis; and protozoa, which can cause toxoplasmosis.

🐦 Fungi have over 100,000 recognized species, 100 of which are infectious to people. Fungi are classified depending on the degree of tissue involvement and mode of entry to the host. Examples of infectious fungi include ringworm (on the skin) and Histoplasmosis (inside internal organs). [*See* Fungi, Pathogenic, *page 59.*]

🐦 Have you ever heard of prions? Called TSEs (which stands for transmissible spongiform encephalopathies) by scientists, prions are a retched family of rare, progressive, neurodegenerative diseases that affect both humans and animals, like Creutzfeldt-Jakob disease and bovine spongiform encephalopathy ("mad cow" disease), for example.

Photodegradation:
Why Sunlight Fades Sofas

🐦 "Photodegradation" is the technical term for color fading.

🐦 Here's the thing: There are light-absorbing color bodies called "chromophores" that are present in dyes.

🐦 The color(s) we see are based on these chemical bonds, and the amount of light that is absorbed in a particular wavelength.

🐦 If your sofa is near a window then it's being hit by the sun's UV (ultraviolet) light rays.

🐦 Ultraviolet light can break down those chemical bonds, which results in the fading of colors on objects like sofas. Essentially, it's a bleaching effect.

🐦 Some objects are far more prone to fading, such as dyed textiles and watercolors, while objects that are more reflective of light tend to be less prone to fading, like a highly polished coffee table.

🐦 Incidentally, there are other sources and types of photodegradation besides UV rays, such as infrared light and visible light.

P

🔖 Oxidation is a common type of photodegradation reaction that is used by some drinking water and wastewater facilities to destroy pollutants. So photodegradation isn't always a bad thing! That being said, one of the main environmental challenges with using plastic products is that they photodegenerate, or fade and crack, rather than biodegrade.

🔖 Small bits of plastic in the ocean—which can kill sea life and birds that inadvertently consume these bits—are called "mermaid's tears."

Pinnipeds

 From the Latin word *pinna*, meaning "fin" or "wing," and *ped*, meaning "foot," pinnipeds encompass a vast and diverse collection of semi-aquatic marine mammals, including walrus (*Odobenidae*), sea lions, fur seals, eared seals (*Otariidae*), and earless seals (*Phocida*).

 The earliest pinniped fossil to date is estimated to be 23 million years old.

 Because pinnipeds have to spend many consecutive months at sea, they sleep belly-up in the water for one minute at a time.

 Despite being known as the crab eater seals (*Lobodon carcinophagus*), pinnipeds don't actually eat crabs. They do, however, have very cool lobed triangular molars with triangular holes that act like a sieve to filter their favorite food, Antarctic krill. And just for the record, crabs don't live in the seas of Antarctica, which is where the crab eater seals reside.

P

 A weddell seal's body is comprised of about 20% blood. The human body is made up of only about 7% blood. And not only do these seals have nearly three times as much blood as humans, they also have double the proportion of red blood cells, which enables them to carry a lot more oxygen and hold their breath for a lot longer than a human can.

 The largest of all seals is the southern elephant seal. Males can be over 20 feet (6 m) long and weigh up to 8,800 pounds (4,000 kg).

🐋 Southern elephant seals are also incredibly impressive. They have been officially documented for staying under water for nearly two hours and diving to depths of over 4,600 feet (1,400 m).

🐋 The northern elephant seal is perhaps best known for being a deep-water nocturnal feeder, and for its ability to dive to depths from 1,000 to 2,600 feet (300 to 800 m).

🐋 Contrary to popular belief, sea lions are pinnipeds. There are seven species of sea lions in the world.

Polynya

🍂 The word "polynya" comes from the Russian word meaning "natural ice hole." Polynyas are areas of open water in the polar icepacks.

🍂 A number of things can lead to the formation of polynyas, including warm currents swirling up to the surface, wind, and coastal or sea bottom features like shoals or underwater sea mountains.

🍂 Polnyas are important breathing stations for marine mammals like beluga and killer whales, walrus, and narwhals trapped by the seasonal icepack, and migratory waterfowl.

🍂 Note: Narwhals are whales with one tusk that look like unicorns.

🍂 Polynyas are a hunting ground for polar bears hoping to prey on one of the other animals the polynyas attract, like walrus.

🍂 Some polynyas reoccur in the same location year after year.

P

FULL SERVICE POLYNYA

Protostars

🍂 Stars are luminous spheres of gas, fueled by nuclear fusion. Protostars are the precursors to stars. This, we know.

🍂 Conventional wisdom states that protostars are formed when clouds of gas and dust, left over from earlier exploded stars, gather together and then (this is the key part) collapse under the gravity of their own weight, creating the star formation. This theory is currently being challenged!

🍂 Recent scientific discoveries suggest that protostars—which are developing stars, akin to being in the embryonic stage—may have more than gravity to thank for their rising-star status.

🍂 Apparently, some yet-to-be-understood energetic process, probably related to magnetic fields, is superheating the surface of the cloud core, thereby pushing the cloud to become a star.

🍂 Be that as it may, in order to have gravity to initiate the necessary nuclear fusion reaction, a protostar must still have at least 8% of the mass of the Sun, or 75 times the mass of Jupiter. But if it does have the right mass, a protostar should achieve its stardom!

🍂 The absence of enough gravity will cause the star to become a "brown dwarf" that glows with heat in the infrared, but never really shines.

P

Purring

🐾 While people tend to interpret purring as a sound of contentment, no one knows for sure why a domestic cat purrs or exactly how this happens.

🐾 There are scientists who suggest that the purr's low frequency vibrations may function as a natural healing mechanism, which would help explain why some cats purr when injured or in pain.

🐾 What makes the purr distinctive from other cat vocalizations, such as growling or meowing, is that the sound is produced during the entire respiratory cycle of inhaling and exhaling, as opposed to other vocalizations like meowing, which are limited to expiration of the breath.

P 🐾 Scientists generally agree that the larynx (voice box), the laryngeal muscles, and a neural oscillator are involved when a cat purrs.

🐾 We used to think purring was produced from blood surging through the inferior vena cava, but research suggests that the intrinsic (internal) laryngeal muscles are the likely source for the purr.

🐾 If a cat suffers laryngeal paralysis, they no longer purr.

🐾 Kittens learn how to purr when they're just a few days old.

🐾 Purring is a possible bonding mechanism between kitten and mother. There's research that suggests when a kitten purrs it's telling her mother: "I'm okay" and "I'm here."

◦❧ In addition to the domestic cat, other species in the *Felidae* family purr, such as bobcats and cheetah, eurasian lynx, and pumas (also known as mountain lions).

◦❧ Big cats such as lions, leopards, jaguars, tigers, snow leopards, and clouded leopards all exhibit purrlike sounds, but studies suggest they don't exhibit true purring.

Q Fever

🐦 A zoonotic disease, meaning people can catch it from animals, Q fever (*Coxiella burnetii*) is a disease that spans the globe.

🐦 The "Q" in "Q fever" stands for "query" because when it was named, the causative agent was a mystery!

🐦 A very robust bacterium, *Coxiella burnetii* can survive for long periods of time in the environment and can even spread by wind and dust.

🐦 Cattle, sheep, and goats are commonly infected with Q fever and may transmit infection to humans when they give birth. Birds and cats can also be carriers, as well as ticks.

🐦 Humans can get Q fever by drinking raw, unpasteurized milk or by inhaling dust or droplets in the air that contain animal feces, blood, or birth products.

🐦 While Q fever can occur throughout the year, cases usually peak in April and May in the U.S., which coincides with many domestic animals' birthing season.

🐦 About 3% of the healthy U.S. population and 10–20% of people in high-risk occupations like veterinarians and farmers have antibodies to *C. burnetii*, which strongly suggests that a person either had the disease or was previously exposed to it.

---⚜---

❧ With symptoms ranging from unnoticeable to severe, complications from Q fever may include endocarditis, encephalitis, pneumonia, hepatitis, and splenomegaly.

❧ And because I know you're wondering two things: The most common symptoms of Q fever mimic the flu, which is why some people have no idea they've ever had it. Yes, it is treatable; and since it's an intercellular bacterial infection, the antibiotics most commonly used are tetracycline and doxycycline.

Quasars

🦢 The word "quasar" was first coined in 1964 by astrophysicist Hong-Yee Chiu as convenient shorthand for the previous nomenclature "quasi-stellar radio sources."

🦢 Initially perplexing to astronomers and astrophysicists, quasars were first identified as powerful radio signals coming from regions where no visible astronomical source was apparent, meaning they didn't see any stars, planets, galaxies … nothing.

🦢 Until very recently, quasars were described as the most distant objects in the universe, which is why we hadn't seen them before. However, we've since detected a gamma ray burst and a galaxy, both of which are farther away.

🦢 Quasars are now believed to be powered by super-massive black holes at the centers of distant galaxies called active galactic nuclei.

🦢 Quasars are the most luminous objects in the universe, with the brightest ones emitting the energy equivalent to two trillion stars!

🦢 A quasar named ULAS J1120+0641 was discovered in 2011 and is currently the farthest quasar from Earth, at 29 billion light-years away.

🦢 Because of the time it takes light to travel, we see the light of quasar ULAS J1120+0641 as it was when the universe was 770 million years old.

Quillworts

◆ Quillworts (genus *Isoëtes*) are non-flowering, fernlike plants.

◆ Frequently mistaken for small, aquatic grasses, quillworts don't have the requisite hollow stems, nodes, and flat leaves.

◆ Quillworts' leaves are hollow and quill-like, hence the name.

◆ The hollow leaves are less than ⅛ inch (1 to 2 mm) wide and broaden to a swollen base where they attach in clusters to a bulblike, underground rhizome, which is a characteristic of all quillwort species except mat-forming quillwort, *I. tegetiformans*. This swollen base also contains male and female sproangia protected by a thin, transparent covering (velum). Thus, quillworts are heterosporous, meaning that they have both male and female spores.

◆ There are upward of 150 species of quillworts. The leaves of black-spored quillwort (*Isoetes melanospora*) are stiff to arching and usually less than 3 inches (8 cm) tall, but generally longer than those of mat-forming quillwort.

◆ The black-spored quillwort is an endangered species on the state and federal lists for both Georgia and South Carolina.

Q

Quitting Smoking Benefits

"Quitting smoking is easy. I've done it a thousand times."

—Mark Twain

- Twenty minutes after your last cigarette, most of the nicotine leaves your brain.
- Eight hours after your last cigarette, your oxygen levels begin to rise.
- Twenty-four hours after your last cigarette, your risk of heart attack begins to go down.
- Two days after your last cigarette, you will be able to smell and taste things better.
- Two weeks after your last cigarette, you have better blood circulation.
- One month after your last cigarette, you breathe better and feel less tired.
- One year after your last cigarette, your risk of heart disease is cut in half.
- Five years after your last cigarette, your risk of dying of lung cancer has been cut in half.
- Five to 15 years after your last cigarette, your risk of stroke is about the same as someone who has never smoked.
- Ten years after your last cigarette, your risk of lung cancer is nearly the same as someone who never smoked; your risk of pancreatic cancer is also reduced.
- Fifteen years after your last cigarette, your risk of heart disease is as low as if you never smoked.
- Pretty cool, right?

Quokka

~&~ The quokka (*Setonix brachyurus*) is a cat-size marsupial commonly described as one of the smallest wallabies. It is the only member of the genus *Setonix*.

~&~ One day after a baby quokka, called a "joey," is born the female immediately mates again, and thus begins the *embryonic diapause*. This means that there's now a new embryo that stays dormant within the mom, like a baby in waiting!

~&~ If the joey in the mom's pouch dies within five months, the embryo in waiting resumes development and is born in 24 to 27 days!

~&~ If the joey lives, the embryo degenerates.

Q

~&~ Dating back to the 1600s, the quokka was one of the first Australian mammals ever seen by Europeans.

~&~ Found primarily in southwest Australia, especially on the offshore islands of Rottnest and Bald, quokkos are listed as "vulnerable" under the Environment Protection and Biodiversity Conservation Act. [*See* Marsupials, *page 139.*]

Radiocarbon Dating

🐛 All organisms absorb carbon from the environment until they die.

🐛 Radiocarbon dating utilizes a special method of measuring the naturally-occurring isotope Carbon-14 (^{14}C) to ascertain a reasonable estimate of something's age, as long as it's less than 50,000 years old.

R

❧ The concentration of ^{14}C in organisms older than 50,000 years is too small to measure, so they can't be dated via ^{14}C.

❧ So how does the measuring thing work? Okay, ^{14}C is produced by cosmic rays in the stratosphere and upper troposphere, and then makes its way to Earth and its inhabitants.

❧ ^{14}C is a radioactive isotope with a half-life of about 5,730 years.

❧ Remember that organisms stop absorbing carbon when they die? This explains why radiocarbon dating works so well: We know after 5,730 years, half of its ^{14}C will have radioactively decayed to nitrogen. So, a simple calculation can estimate the age, based upon the remaining concentration of ^{14}C in the organism. How cool is that?

❧ Equally cool, measuring the ^{14}C concentrations in old tree rings can test radiocarbon dating: While a tree may live for hundreds, even thousands, of years, each ring of a tree absorbs carbon only during the year in which it grows.

Raining: Fish and Frogs

❧ The belief that it can rain fish and frogs dates back to ancient times. And given the right conditions, it sure can *look* that way!

❧ Tornadoes and hurricanes are powerful enough to suck up a school of fish or frogs and then "rain" them miles away.

❧ Tornadic waterspouts—which are basically tornadoes that form over land and travel over the water reaching speeds of up to 100 mph (160 km/h)—can do the same thing.

❧ In fact, small ponds (water, fish, frogs and all) can be easily sucked up by a strong enough passing tornado, or pulled into a tornadic waterspout's swirling updraft.

❧ Once a swirling vortex starts to lose enough energy (whether it's a tornado or tornadic waterspout), it will "rain" back down what had been sucked up—including fish and frogs!

❧ But does it still rain fish or frogs in "modern times"? In a word: Yes!

❧ In 1873, it rained frogs in Kansas City and there were no swamps or other bodies of water in the vicinity.

❧ In 1882, it hailed frogs in Dubuque, Iowa. Scientists theorize the frogs were picked up by a powerful updraft and frozen into hail in the cold air above Earth's surface.

R

And more recently, on October 23, 1947, it rained fish in Marksville, Louisiana; on June 7, 2005, thousands of frogs rained on Odzaci, a small town in northwestern Serbia; and, in February 2010, hundreds of spangled perch rained down upon the village of Lajamanu, Australia. And the list goes on!

One more really cool fact: The reports are always fish *or* frogs. Not fish *and* frogs. And that makes perfect sense because animals of similar size and weight would naturally be deposited together.

Raspberries

- There are over 200 species of raspberries in the world.

- Red raspberries are believed to have originated in Turkey and spread throughout Mediterranean Europe.

- The Romans took raspberries throughout their empire, including to Britain.

- In the 16th century, people in England started cultivating raspberries.

- When explorers and settlers arrived in North America, they were pleased to discover the black raspberry. However, they contained more seeds and weren't as sweet as the red raspberry of Europe, so the settlers planted red raspberry seeds that they brought with them.

- The British starting shipping red raspberry plants to New York in 1771.

- About 90% of all raspberries in the U.S. are grown in Oregon, Washington, and California.

- Raspberries can come in many colors, including red, black, purple, and golden/yellow. It's their colors, their pigment-causing flavonoids, that reveal which antioxidants they likely contain. Anthocyanins, for example, appear as red, purple, or blue.

R

꩜ In addition to having a hollow central core, raspberries are actually made up of many connecting, individual sections of fruit, each with its own seed.

꩜ Raspberries have an exceptional amount of antioxidants, like quercetin, anthocyanins, catechins, pelargonidins, kaempferol, gallic acid, and cyanides—color dependent, of course!

꩜ According to a 2011 study published in the *Journal of Agriculture and Food Chemistry*, regular consumption of flavonoid-rich berries such as raspberries may help prevent and/or moderate chronic diseases.

Reefs

 Coral reefs are called the rain forests of the sea and enjoy an extraordinarily biologically diverse ecosystem, providing food and shelter to millions of species like fish, sponges, sea anemones, bryozoans, worms, sea stars, crustaceans, and snails, to name a few.

 Reefs first came into existence about two billion years ago. The first reefs were built by colonies of calcareous algae, not corals. This was during the mid to late Precambrian era.

 It was during the Mesozoic era 65 to 245 million years ago when hard corals began to flourish, but they got wiped out.

R The corals of Tertiary period, 2 to 65 million years ago, are the most similar to species we have today. But they were mostly wiped out as well.

 The coral reefs that exist today are less than 10,000 years old.

 Most commonly found in warm, shallow locations in tropical waters, coral reefs can be found throughout the tropical and subtropical western Atlantic and Indo-Pacific oceans, although they can also be found in deep water, particularly near volcanic islands.

 Charles Darwin first proposed the basic coral reef classification that's still used today!

● The three primary categories of coral reefs are:

  ◦ Atoll reefs, which are oceanic, annular, more or less circular reef systems that wrap around a lagoon without a central island. The largest atoll is named "kwajalein," and they're located in the Indo-Pacific.
  ◦ Barrier reefs, which have a deep channel paralleling, but separate from, the shore of an island or mainland. The Great Barrier Reef off northern Australia in the Indo-Pacific is the largest barrier reef in the world.
  ◦ Fringing reefs, which are either physically attached or fairly close to a shore, and have an entirely shallow lagoon or no lagoon at all. Commonly found in the South Pacific Hawaiian islands and parts of the Caribbean, fringing reefs are also found in locations such as off the coast of Eilat, Israel.

● Today, many of our world's reefs have become severely damaged or destroyed by human activities like water pollution, overfishing and destructive fishing practices, global climate change, and ship groundings.

Reptiles

ℜ

> ❧ According to the CDC, an estimated 3% of households in the United States own at least one reptile.

> ❧ Children under five and people with weak or compromised immune systems should avoid contact with lizards, snakes, turtles, and tortoises, as they're associated with the bacterial disease salmonellosis, which infects about 70,000 people in the U.S. every year.

> ❧ Reptiles are amniotes, meaning their eggs are protected from desiccation (drying out) thanks to an extra membrane, called the amnion.

> ❧ Not all reptiles lay eggs; some instead give birth to live young.

> ❧ Oviporous reptiles lay eggs with a shell from which their babies emerge.

> ❧ Ovoviviporous reptiles carry eggs internally, but upon delivery, the shell has so thinned that only the mucous membrane remains from which the babies emerge.

> ❧ In viviparous reptiles, no internal egg structure is present at any time during the development of the babies.

> ❧ Unlike people whose lower jaw is a single bone, a reptile's lower jaw is comprised of several bones, affording it far greater bite mobility.

———————————— ⚜ ————————————

🐦 All reptiles have (or did have, in their evolutionary history) horny epidermal (skin) scales made of the protein keratin (which is what human nails and hair are made of); paired limbs (arms and legs); feet with five toes each; lungs for breathing; and, have a three- or four-chambered heart.

🐦 The teeth of reptiles are all basically cone-shaped and vary only slightly in size.

🐦 Subdivided into four main groups, reptiles include crocodilians like alligators, caimans, crocodiles, gharials; squamates, like lizards, snakes; tuataras, like *Sphenodon punctatus* and *Sphenodon guntheri* (both very rare); and turtles, of which there are nearly 300 species. [*See* Tuataras, *page 233.*]

Robots

❧ A writer named Karel Capek first coined the term "robot" in his 1920 play *Rossums Universal Robots*. Although in his play, robots were more biological and more like clones or androids as opposed to the more conventional metal and mechanical robots.

❧ In the *Star Trek* episode "Requiem for Methuselah," the android's name is Rayna Capek. Leonardo da Vinci sketched plans for a humanoid robot back in 1495!

❧ In his 1942 short story "Runaround," author Isaac Asimov devised the Three Laws of Robotics:

R

      ✑ A robot may not injure a human being or, through inaction, allow a human being to come to harm.

      ✑ A robot must obey the orders given to it by human beings, except where such orders would conflict with the First Law.

      ✑ A robot must protect its own existence as long as such protection does not conflict with the First or Second laws.

❧ In 1932, the first true robot toy was produced in Japan—a wind-up toy that could walk. Called "Lilliput," the toy stood just under 6 inches (15 cm) tall.

Rogue Waves

🌊 Rogue waves are special kinds of waves that are distinguished by an instant, singular, and unexpected wave profile with an extraordinarily large and steep crest or trough. This means they're crazy huge, particularly considering that they occur in otherwise-calm weather conditions.

🌊 Rogue waves are sometimes called freak waves. Although less common, they're also called monster waves, killer waves, extreme waves, and abnormal waves.

🌊 Rogue waves can happen in lakes as well as in the ocean.

🌊 It's only recently that rogue waves been officially recognized as a real wave. They had been largely dismissed as an old sailor's tale!

🌊 Because of their tremendous size and sudden appearance—they're practically a vertical wall of water!—this natural phenomenon is severely hazardous to merchant marines, offshore platforms, naval fleets, as well as other sea-going ventures. How big are we talking? Rogue waves can be upward of 115 feet (35 m), which is taller than a ten-story building … Yikes!

🌊 While it's hard to fathom, rogue waves happen in the Great Lakes in North America! In fact, it's now believed that rogue waves caused the sinking of the SS *Edmund Fitzgerald*, a freighter, in November of 1975. At the time, it was the largest ship on the Great Lakes.

R

❧ The SS *Edmund Fitzgerald* sank so quickly that no distress signals were received.

❧ When it was located on the bottom of the lake later that month, the SS *Edmund Fitzgerald* was found to have been broken in two. One theory is that the ship was hit by the Three Sisters, the name for the phenomenon of rogue waves occurring in sets as opposed to a single, massive wave.

❧ In 1976, "The Wreck of the Edmund Fitzgerald" was a hit song by Gordon Lightfoot.

Roller Coasters: History and Fun Facts

❧ The original "ancestor" of the roller coaster can be traced to 15th-century Russia, where they had a gravity sled ride called Russian Mountains!

❧ In 1827, the Mauch Chunk Switchback Railroad in Summit Hill, Pennsylvania, built a track 18 miles (30 km) down a mountain to transport coal. In 1873, it became a scenic, albeit bumpy, pleasure ride. It remained in operation until 1938.

❧ Called the "father of gravity," La Marcus Thompson built the Switchback Railway at Coney Island in Brooklyn, New York, in 1884.

R

❧ La Marcus Thompson also holds several patents, including U.S. Patent 310,966 (1885) for "roller coaster structure" and U.S. Patent 1,102,821 (1914) for "signaling device for racing coasters."

❧ One of the first roller coasters, *Les Montagnes Russes à Belleville*—meaning "Russian Mountains of Belleville"—was opened in France in 1817. A carved groove attached the coaster's axle to the track.

❧ One of the first "high-speed" coasters was Drop-the-Dip, at Coney Island in Brooklyn, New York, in 1907. Thankfully, lap restraints had started to be used by then.

 The first coaster to completely invert passengers was the Corkscrew (1975), located at Knott's Berry Farm, in Buena Park, California.

 In 1959, Disneyland (in California) introduced the first tubular steel roller coaster. Prior, they were wooden.

 Today, to test a roller coaster's safety the first "riders" are sandbags or dummies. Then, engineers (and sometimes park workers!) take a test ride. Um, don't think I'd volunteer for that!

Roller Coasters:
Physics

෴ The reason why roller coasters stay on their tracks—and why people can hang upside down without falling out—is a matter of physics, or, more specifically, of energy, inertia, and gravity.

෴ As roller coasters don't use engines, the climb up the first hill is accomplished by a lift or cable that pulls it up. This builds up the potential energy that's used to go screaming down the hill, with the help of gravity. Then, all of that stored energy is released as kinetic energy, which is what propels the coaster up the next hill.

෴ As the coaster travels up and down hills, its motion is constantly shifting between potential and kinetic energy, and the higher any given hill, the faster it'll swoosh back down.

R

෴ According to Newton's first law of motion, "An object in motion tends to stay in motion, unless another force acts against it." So, the wheels along the track are forces that work to slow down the roller coaster, which is why the hills tend to be lower towards the end of the ride.

෴ Centripetal force, which is the sensation of being thrust against the outside of the car as you whirl around a turn, helps keep you in your seat.

෴ When doing the upside down loop-de-loop, it's inertia that keeps you in your seat. However, the loop must be elliptical rather than a perfect circle, otherwise the centripetal G-force would be too strong for both safety and comfort! [*See* G-Force, *page 167.*]

Saturn

❧ A gas giant and the second-largest planet in our solar system, Saturn is the sixth planet with respect to the Sun and is perhaps best known for its stunning rings (although, Jupiter, Uranus, and Neptune also have rings).

❧ Comprised mostly of hydrogen and helium, the two lightest elements, Saturn is the lightest and least dense planet that we know of in the universe.

❧ Saturn is so light, it would float on water.

❧ Saturn's seven rings stretch 169,800 miles (273,266 km) wide. However, the rings are remarkably thin. If you turned the rings on their side, they'd only be about 300 feet (91.44 m), which means they would fit between the goal posts on a football field!

❧ Comprised of ice, dust, and rocks, the beautiful concentric rings' spacing is caused by the gravity of Saturn's 53 official and nine provisional, unofficial moons.

❧ Perhaps the most well known of Saturn's moons is Titan, which is larger than the planet Mercury. In fact, the only larger moon in our solar system is Jupiter's Ganymede.

❧ Saturn is 840 million (1.35 billion km) miles from the Sun and has a revolution period of 29.5 Earth years.

➥ What's in a name? The Roman god of agriculture, Saturn was the son of Uranus and the father of Jupiter. But, as the saying goes, "What goes around comes around." While Saturn successfully overthrew his father Uranus to become king of the gods, Saturn was then overthrown by his own son, Jupiter!

Sea Lions

🦭 From the Latin word *pinna*, meaning "fin" or "wing," and *ped*, meaning "foot," sea lions are pinnipeds. [*See* Pinnipeds, *page 183.*]

🦭 Killer whales—which are actually dolphins—and sharks are sea lions' only natural predators.

🦭 Found in every body of water except the Atlantic Ocean, there are seven species of sea lions:

- California sea lions are believed to be the most intelligent of all the sea lions.
- Steller sea lions are the largest of the sea lions.
- Australian sea lions are an endangered species, due primarily to overhunting and a continuing loss of habitat. Today, there are less than 10,000 alive in the world and it is illegal to harm or kill one under the authority National Parks and Wildlife Act of 1972.
- Galapagos sea lions are known for their exceptionally playful and social nature, their unusual pointy noses, and, compared to the other six species, they sound most like barking dogs.
- New Zealand sea lions, formerly known as the hooker's sea lion, are also declining in numbers and rank as the second-rarest sea lion in the world.
- The South American sea lion tend not to cluster together as other sea Lions do. Rather, they prefer some personal space between them. This may be due in part to the warm climates in which they live, like Chile, Peru, Uruguay, and Argentina. The technical name for the South American sea lion is the

S

Patagonian sea lion ... but for reasons unknown, they don't seem to be called this.

∾ Japanese sea lions were officially pronounced extinct on the IUCN Red List of Endangered Species in 1990. The conservation efforts for the Japanese sea lion came too late. Their downfall was primarily due to extensive and unregulated commercial harvesting for their beautiful skin that was used for bags and apparel, meat and blubber, and medicinal uses perpetrated by the predators—known as humans.

Silver

- From the Latin word *argentums*, meaning "shining," the chemical symbol for silver is Ag.

- Silver has the highest level of electrical conductivity of all metals—even higher than copper!

- … But copper is cheaper.

- In fact, silver literally defines electrical conductivity. Yes, silver sets the standard by which all other metals are compared. On a scale of 0 to 100, silver ranks 100, copper ranks 97, and gold ranks 76.

- Sterling silver is not pure silver; it's an alloy, meaning a combination of metals—usually about 92.5% silver by weight and 7.5% copper and/or other metals.

- Having an incredible degree of optical reflectivity, a silver mirror can reflect about 95% of the visible light spectrum (which is about the same as gold).

- Silver has the highest level of thermal (heat) conductivity of all other metals, which is why it's used in solar panels.

- A single ounce (28 g) of silver can be stretched into about 8,000 feet (2,438 m) of wire.

◈ Silver is mentioned multiple times in the book of Genesis: "And Abraham was very rich in cattle, in silver, and in gold" (Genesis 13:2).

◈ One of the most difficult words with which to find a rhyming word in the English language is silver. Other examples of difficult-to-rhyme words include orange, purple, wolf, angst, gulf, and ninth.

Smallpox Vaccine History

🕊 Physician and scientist, Dr. Edward Jenner is often referred to as the father of immunology, and the pioneer of the smallpox vaccine.

🕊 However, Lady Mary Wortley Montagu, a well-respected 18th-century writer, was an incredibly significant—albeit generally overlooked—contributor to the eradication of smallpox.

🕊 It wasn't until after Dr. Edward Jenner read Lady Montagu's report that he published his own report on his smallpox vaccine! Lady Montagu's report, published 80 years earlier, documented how Turkish women would apply cowpox to their children's wounds to stave off further infection,

🕊 Montagu had contracted smallpox herself in 1715, which disfigured her beautiful face. She became determined that no one else should get it.

🕊 Cowpox is a very mild disease, which primarily causes a few blisters on the hands. However, it's close enough to trigger a protective immune response against smallpox.

🕊 While we're on the subject of giving credit where credit is due:

🕊 One hundred years before Jenner, the Chinese realized that if they scraped smallpox scab samples off people with a mild version of the disease, and then purposely infected healthy people with it by inserting it up their nose or putting it on the surface of their skin,

the person would not get the crippling or deadly version. And later, the Chinese made a pill version using powdered smallpox scabs. This deliberate act of inoculation via an infectious agent is called "variolation."

- By 1721, Dr. Zabdiel Boylston was using variolation in the New World, successfully stopping a smallpox epidemic in Boston, Massachusetts.

- In 1777, George Washington had all his soldiers variolated before beginning new military operations.

- In 1796, Dr. Jenner extracted fluid from a cow pock pustule on a milkmaid and used it to inoculate a healthy eight-year-old boy named James Phipps. Six weeks later, Jenner purposively tried to infect the boy with smallpox but he did not contract it. This led to the development of a smallpox vaccine, which induced an immune response, without infecting someone with the actual disease.

- Thus, it should come as no surprise that the word "variolation" comes from the Latin word *variola*, meaning "smallpox."

- And, the word "vaccine" comes from the Latin word *vaca*, which means "cow"!

Solar Energy

- With the ability to convert sunlight directly into electricity, solar cells are called photovoltaic cells, or PV devices.

- During an expedition to Africa back in the 1830s, the British astronomer John Herschel used a solar thermal collector box—a device that absorbs sunlight to collect heat—to cook food!

- The International Space Station uses solar panels to help generate power.

- The largest solar power plant in the world is located in the Mojave Desert.

S

- Solar energy systems don't produce air pollutants or carbon dioxide, making solar energy an incredibly clean source of energy. However, as the amount of sunlight that arrives at Earth's surface is not constant (it varies depending on location, time of day, time of year, and weather conditions), there are still challenges to be overcome with its use.

- In 2010, solar thermal-power generating units were the main source of electricity at 13 power plants in the United States: 11 in California, one in Arizona, and one in Nevada.

Sun

ꙮ Born in a vast cloud of gas and dust around five billion years ago, the Sun is basically a giant ball of hydrogen and helium gases. More specifically, approximately 74% of the Sun's mass is made up of hydrogen; 24% is helium, and the remaining heavier elements are oxygen, carbon, iron, and neon, which make up the remaining 2%.

ꙮ Sunlight can reach a depth of around 262 feet (80 m) in the ocean.

ꙮ At the core, the temperature of our Sun is around 27 million °F (15 million °C).

ꙮ The Sun's surface temperature is around 9,941°F (5,505°C).

ꙮ In the Sun's core, hydrogen atoms combine to form helium atoms. This process, called fusion, gives off radiant energy. [*See* Nuclear Fusion, *page 163.*]

ꙮ The Sun is about 865,000 miles (1,392,000 km) wide, which means that the Sun's diameter is about 110 times wider than Earth's.

ꙮ It takes about eight minutes for the light from the Sun to reach Earth.

S

Sunspots

꧁ ༺ ༻ ꧂

꧁ Coming and going on a regular basis, sunspots appear darker because they're cooler in temperature ... relatively speaking. Sunspots are approximately 2,732°F (1,500°C) cooler than the rest of the Sun.

꧁ Sunspots tend to appear in pairs.

꧁ Some sunspots are bigger in size than Earth.

꧁ The dark centers of sunspots are called the "umbra"; the surrounding lighter colored regions are called the "penumbra."

꧁ Generally, sunspots increase in intensity and then decrease over a period of 11 years. This process is called the Saros cycle.

꧁ Sunspots have more magnetic activity then the rest of the Sun.

꧁ Scientists generally agree that sunspots and their accelerated charged particles (which are primarily electrons) are the catalyst, if not the cause, for solar flares.

S

Syzygy
(One Word, So Many Meanings!)

🐦 From the Latin word *syzygia*, meaning "a conjunction," the word "syzygy" has a variety of meanings, ranging from the astronomical, where it is recognized as English in 1847, to zoological!

🐦 In astrological terms, the word is defined as an alignment of three or more celestial bodies in the same gravitational system along a straight line. For example, in 1982 there was a syzygy, where all eight planets (plus the now-dwarf planet Pluto) all lined up in a straight line with the Sun.

🐦 *[Bonus factoid for my fellow uber-nerds: "Syzygy" was the name of a 1996 X-Files episode, where the alignment of Mercury, Mars, and Uranus occurred at the same time and several murder cases occur in a small town.]*

S

🐦 Biological syzygia refers to the pairing of chromosomes in meiosis.

🐦 Cable: On the April 18, 2007 episode of *The Colbert Report*, Stephen Colbert used "syzygy" in its poetic sense—after synecdoche and metonymy—as part of an all-in-good-fun "threat" made against actor Sean Penn, in preparation for the next night's "Metaphor off" between the two.

🐦 The term has also been used in graphic novels. Syzygy is a key character in the space opera *Dreadstar*, premier title of *Epic Comics*; Syzygy is a webcomic by Alicorn; and, for a brief period in the Dr. Strange comics, Strange invoked the syzygy of all planets to tap into "catastrophe magic."

Computational: A name given to a virtual reality grid-operating system for PC clusters. A PC cluster is a way of combining a bunch of PCs to get super-computing power levels—since virtual reality images require a ton of computations!

In mathematical terms, syzygia is the relation between the generators of a module.

In medical terms, syzygia is the fusion of some or all of the organs without the loss of identity by either.

In psychological terms, syzygia has been used by psychologist Carl Jung to denote an archetypal union of the conscious and unconscious mind without a loss of identity for either.

In zoological terms, syzygia refers to the end-to-end or lateral association of two for the purpose of asexual genetic material exchange.

Telegraph

❧ Existing since ancient times, telegraphs—which, by definition requires there be no physical exchange of message-bearing objects or creatures (including pigeons)—are simply forms of long-distance communication, e.g., smoke signals, reflected lights, flags, and beacons.

O-M-G-! L-O-L-?! WHAT IN THE WORLD ARE THEY TALKING ABOUT? THIS MUST BE BROKEN.

T

❧ The word "telegraph" comes from the Greek word *tele*, meaning "far," and *graphein*, meaning "writing."

❧ In 1794, Claude Chappe invented a handheld, flag signal-based alphabet (a semaphore) that could be used over long distances. However, a clear line of sight was required between users.

❧ Napoleon Bonaparte was a fan of the semaphore flag system, as was the United States during the Peninsular War (c. 1807 to 1814).

❧ In 1825 British inventor William Sturgeon invented the electromagnet, and in 1830 Joseph Henry demonstrated the potential of long-distance communication via an electric magnet. Based on the work of these two inventors, William Cooke and Charles Wheaton patented the telegraph. Contrary to popular belief, it was not Samuel Morse.

❧ Samuel Morse was the first person to successfully apply and market the commercial use of the telegraph in 1837. However, it was Cook and Wheaton who first demonstrated the telegraph's *potential* for commercial application. [*See* International Morse Code, *page 89.*]

Thunder

🌀 Contrary to popular belief, the thunderclap that follows lightning bolts doesn't come from the bolt itself.

🌀 The grumbles and growls heard during thunderstorms actually come from the rapid expansion of the air surrounding the lightning bolt.

🌀 Vertical lightning is often heard in one long rumble. [*See* Lightning, *page 132.*]

🌀 The shock waves nearer to the ground reach your ear first, followed by the crashing of the shock waves from higher up.

T 🌀 When a lightning bolt is forked, sometimes resembling tree branches, it sounds different. The shock waves from the various lightning branches bounce off each other, as well as off of low hanging clouds and nearby hills, and that's what creates the series of lower, continuous grumbles of thunder!

🌀 The sound of thunder isn't exclusive to thunderstorms; thunder can occur during snowstorms, albeit rarely.

🌀 Lightning doesn't always create thunder. [*See* Lightning, *page 133.*]

🌀 To estimate approximately how close lightning may be to you, count the seconds between the flash and the thunderclap. Each second represents 984.25 feet (about 300 m).

Titan Arum

(a.k.a. Corpse Flower)

🌿 An enormous plant, the titan arum (*Amorphophallus titanium*) plant can reach heights of 7 to 12 feet (2 to 4 m) and weigh as much as 170 pounds (70 kg)!

🌿 Similar to the *Rafflesia arnoldii*, the titan arum has a dreadful smell, best described as being akin to the smell of a decomposing mammal, hence the nickname "corpse flower." [*See* Flowers, *page 52.*]

🌿 Indigenous to the rain forests of Sumatra in Indonsia, the titan arum is called a *bunga bangkai—bunga* meaning "flower," and *bangkai* meaning "corpse" or "cadaver."

🌿 The bloom of the titan arum is not a single flower. It's actually an *inflorescence*, which is a cluster of many tiny flowers arranged on a stem.

🌿 In fact, the titan arum has the largest unbranched inflorescence of all flowering plants known in the world.

🌿 It was naturalist Sir David Attenborough who invented the name "titan arum" for his BBC series *The Flowering Life of Plants*. Why the new name? Calling it by its proper name, *Amorphophallus*, struck him as inappropriate for his family-friendly series.

T

Tornadoes

❧ Tornadoes are spawned within thunderstorms when the cloud mass overhead starts rotating and begins to pull warm air in from below.

❧ The strongest tornadoes are produced by thunderstorms called supercells, which differ from other thunderstorms because they contain a circulating air mass called a mesocyclone.

❧ From 1971 to 2007, tornadoes in the U.S. were graded on the Fujita scale, a six-point scale (F0 to F5) that classifies tornadoes based on the damage they cause to structures and vegetation.

❧ The Fujita scale is not a measure of wind speed, although it contains estimated ranges for the wind speed necessary to cause the observed damage.

❧ In 2007, the U.S. switched to the enhanced Fujita scale. While still a six-point scale (EF0 to EF5), estimated wind speeds needed to cause certain levels of damage are lowered on the enhanced scale. It also has a more extensive scoring system, with many different structure types and degrees of damage to improve consistency and standardization of tornado classification.

❧ Waterspouts—and their counterpart landspouts, or "dust tube tornadoes"—are formed by a different mechanism than most tornadoes. They are formed in areas of converging moist winds when the convection, or heat rising, caused by a newly forming cloud reaches from the ground up to the cloud and begins to spin.

Tsunamis

🦆 A tsunami is not just one wave, it's actually a series of waves that arrive every 10 to 60 minutes.

🦆 That first enormous wave is not usually the biggest.

🦆 Underwater earthquakes, landslides, or volcanic eruptions can cause tsunamis. While rare, a tsunami can be generated by a giant meteor slamming into the ocean.

🦆 When a tsunami is traveling across the ocean, it doesn't look like that giant wall of water we all imagine.

🦆 The top of the water may only be a couple of inches higher than usual!

T

🦆 The giant waves are mostly happening *under* the surface of the water and can be miles deep.

🦆 Tsunamis are sometimes erroneously called "tidal waves," because initially when out at sea, they don't look like waves; they look like a normal rising tide.

🦆 A tsunami can travel across the ocean upward of 500 mph (805 km/h)! That's jet airplane speed.

🦆 Tsunamis can occur at any time of day or night, under any and all weather conditions, and in all seasons.

 The phenomenon called a "drawback" happens when, upon approaching land, the water is so shallow that a vacuum effect occurs, temporarily sucking back the coastal waters several hundred feet. If you ever see that happen, RUN to high ground as fast as you can.

 The shallower the water at the coastline is, the taller the wave becomes.

 About 80% of tsunamis happen within the Pacific Ocean's "Ring of Fire," encircling Alaska, Sea of Japan, Pacific Ocean, and South America.

 In the center of the Pacific Ocean, there's one really big tectonic plate that's moving in a northwest direction. That plate inevitably bumps up against a landmass and keeps pushing and pushing. Because it's rock against rock, it gets stuck. Thus, the only thing it can do is break, which causes an earthquake under the ocean. This, in turn, lifts the ocean, which causes a huge displacement of water. A tsunami is born.

Tuatara

🐦 The tuatara is the sole remaining beak-headed reptile of the *Rhynchocephalia* order. There are only two living species of tuatara in existence in the world: The *Sphenodon punctatus* and the *S. guntheri*.

🐦 Also one of the most unevolved animals today, tuatara have successfully sidestepped extinction for at least 200 million years!

🐦 Tuatara can be found only in New Zealand.

🐦 First evolving during the Triassic world of Gondwana, tuatara survived the Cretaceous period when the dinosaurs died out! (Hmm, perhaps saying they "sidestepped extinction" is a bit of an understatement!)

🐦 Tuatara take between ten and 20 years to reach sexual maturity, and they have the longest reproductive cycle of any living reptile. Female tuatara lay between five and 18 eggs, and they only do so once every four years! Fortunately, tuatara live 60 to 100 years.

🐦 Tuatara have an incubation period of 12 to 15 months. They also experience the unusual phenomenon where embryonic development stops during the winter, so a hatchling tuatara would have been conceived over two years prior.

🐦 Ground temperature determines the gender of the hatchlings.

T

🦎 Male tuatara have no external sex organs. Copulation is achieved by a meeting of the cloacal regions in what is known as a "cloacal kiss." The same cloacal region is also used for urine and feces excretion.

🦎 Unlike every other living toothed reptile, the tuatara's teeth have no roots or sockets. Instead, their teeth are fused to the jawbone. This is called having an "acrodont" tooth structure.

🦎 And yes, I know they look a lot like lizards, but they are not lizards!

Turing, Alan

🕊 Alan Turing, a British mathematician and logician, is considered one of the fathers of computer science and artificial intelligence, as well as one of the most accomplished code-breakers of all time.

🕊 In 1936, Turing published a description of a mathematical and logical concept—now called a "Turing machine"—that's essentially the basis for all modern computers. Yes, he invented the concept of computers!

🕊 While working at the Government Code and Cipher School (the British code-breaking agency) during World War II, Turing successfully broke the German Navy Enigma code system. This gave the Allies a critical advantage in the war of the Atlantic against German U-boats.

🕊 Turing was awarded the Order of the British Empire (OBE) in 1945 for his wartime services, but much of his work remained secret for many years.

🕊 In fact, several of Turing's papers on code-breaking were considered so important by the British government, they weren't declassified and released to the public until April 2012—some 70 years after they were written!

🕊 In 1950, Turing published a paper describing the Turing test, a method that determines whether a computer possesses intelligence!

T

❧ The basic idea of a Turing test is this: If a person cannot tell whether they are exchanging text messages with a person or a computer program, then the computer has "passed" the Turing test and could be considered intelligent.

❧ Each year since 1966, the Association for Computing Machinery has awarded the Turing Award in recognition of technical contributions to the computing community. The Turing Award is considered the computer science equivalent of the Nobel Prize.

Twinkle, Twinkle, Little, Big, and Spinning Stars

- Stars twinkle and shine because the gas inside them is so hot that nuclear fusion takes place.

- Nuclear fusion occurs when two atoms fuse and form a different kind of atom. This process emits a lot of energy that we can see as light.

- Coming in every color of the rainbow, a star's color is one of the key indicators used to determine its temperature.

- Stars are responsible for the manufacture and distribution of heavy elements, such as carbon, nitrogen, and oxygen, which impact the characteristics of the planetary systems that may develop near them.

- The fastest spinning star ever discovered is called VFTS 102, which rotates at a dizzying one million mph (1.6 million km/h)! Any faster and it would likely be torn apart due to centrifugal forces!

- By far the most numerous stars in the universe are red dwarfs, which have life spans of tens of billions of years.

- Red dwarfs are the smallest known stars, emitting only 0.01% of energy as compared to the Sun.

❧ Hypergiants are the most massive stars but have relatively short life spans of only a few million years. These stars do way more than twinkle: They burn BIG, emitting hundreds of thousands of times more energy than the Sun! Considered rare, there are only a handful of these super-massive stars in the entire Milky Way.

❧ Generally speaking, the larger a star, the shorter its life.

Ultraviolet (UV) Rays

꙳ Ultraviolet (UV) rays are a part of sunlight that is an invisible form of radiation.

꙳ UV rays can penetrate and change the structure of skin cells.

꙳ There are three types of UV rays: Ultraviolet A (UVA), ultraviolet B (UVB), and ultraviolet C (UVC).

꙳ UVA is the most abundant source of solar radiation at Earth's surface and penetrates beyond the top layer of human skin.

꙳ Scientists believe that UVA radiation can cause damage to connective tissue and increase a person's risk for developing skin cancer.

꙳ UVB rays penetrate less deeply into skin, but can still cause some forms of skin cancer.

꙳ When exposed to UVB rays at an index greater than three, the body produces vitamin D (more specifically, vitamin D3).

꙳ Ultraviolet C (UVC) rays are not believed to pose a risk to people because they're absorbed by Earth's atmosphere.

꙳ Snow and light-colored sand reflect UV light and increase the risk of sunburn.

U

🍂 SPF stands for sun protection factor and specifically refers to the amount of time that a person will be protected from a burn.

🍂 An SPF of 15 will allow a person to stay out in the sun 15 times longer than they normally would be able to stay without burning.

🍂 The SPF rating applies to UVB exposure only, not UVA.

🍂 For UVA protection, look for products containing mexoryl, parsol 1789, titanium dioxide, zinc oxide, or avobenzone.

Ungulates

🐦 Ever heard the term "ungulate"? Ungulates are mammals, most of whom walk on the tips of their hoofed toes. They account for the vast majority of large herbivores currently walking—or rather, tiptoeing!—on Earth.

🐦 The Smithsonian Institution presently recognizes 257 modern ungulate species, five of which have become extinct over the past 300 years, primarily because of people.

🐦 Formerly considered a single order of mammals, ungulata "can" be divided into six subgroups:

- **Artiodactyla:** Even-toed ungulates who bear their weight about equally on the third and fourth toe. These include pigs, peccaries, hippopotamuses, camels, chevrotains (mouse deer), deer, giraffes, pronghorn, antelope, sheep, goats, and cattle.
- **Perissodactyla:** Odd-toed ungulates who bear their weight on their third toe, like horses, tapirs, and rhinoceroses.
- **Tubulidentata:** Aardvarks
- **Hyracoideam:** Hyraxes
- **Sirenia:** Dugongs and manatees
- **Proboscidea:** Elephants

U

🐦 Why did I write "can" in quotes? Well, the different groups of ungulates aren't as closely related as once believed, based upon DNA analysis.

🐌 As a result, ungulates—as a grouping—have no taxonomic significance in some scientific circles.

🐌 That said, they're technically categorized as ungulates. However, the categories are continuing to be redefined—and garnering scientific consensus can be a long, arduous process.

Universe

🐦 The universe is believed to be 13.75 billion years old.

🐦 While estimates vary, the number of galaxies in the universe is believed to exceed 100 billion; galaxies with less than a billion stars are considered "small galaxies."

🐦 Scientists' best estimate of the diameter of the universe is that it spans over 150 billion light-years … and it's still expanding.

🐦 The universe's expansion is also accelerating, which implies the existence of a form of matter with a strong negative pressure, called "dark energy."

🐦 If dark energy does in fact play a significant role in the evolution of the universe, then in all likelihood the universe will continue to expand forever.

🐦 According to NASA's Wilkinson Microwave Anisotropy Probe (WMAP), we now know that the universe is flat (with only a 0.5% margin of error).

🐦 Yes, the universe is flat! Here's the explanation, in brief, from NASA:

> 🐦 The WMPA spacecraft can measure the basic parameters of the Big Bang Theory—including the geometry of the universe.
> 🐦 If the universe were flat, the brightest spots we see in a "microwave wavelength photograph/image" of the universe (called microwave background fluctuations) would be about one degree across. If the universe were closed, the brightest spots would be greater than one degree across.

- The brightest spots are about one degree across, and these findings have been confirmed by DASI, Maxima, MAT/TOCO, and Boomerang!
- So, it's flat!

Untruths, Truthiness, and Just Plain False

❧ Popularized by Stephen Colbert, the term "truthiness" refers to wishing or believing something to be true, despite the facts. Consider this example of truthiness: There is a theory that we use only 10% of our brain, which is entirely untrue. This concept persists, due in no small part to our desire to believe that we have 90% of untapped intelligence potential … if we could only access it.

❧ It's untrue that you have to drink eight 8-ounce (225-ml) glasses of water, under normal conditions. One needs to have that much *fluid* each day, but it needn't come from water. Fruits, vegetables, and pretty much everything we eat or drink contains fluids.

❧ The belief that you lose most of your body heat through your head is untrue. We really only lose 7–10% from our head! Even the temporary increase in blood flow to the brain when exercising quickly reverts back to normal once a body's natural cooling processes, like sweating, are engaged.

❧ The belief that going outside with wet hair when it's freezing cold outside will make you catch a cold isn't true; only viruses cause colds. That said, being outside in frigid weather for too long can lower your body temperature, which may stress your immune system, making you more susceptible if you come in contact with a virus.

U

❧ It's just plain false that the flu shot can give you the flu. Only the benign outer shell of the virus is used in the shot, which your immune system perceives as harmful; thus, it produces disease-fighting antibodies and stores the "attack plan" in a type of virus recognition software for future use. Post-vaccination, your immune system needs about two weeks to "learn" what to do. During that time, some people's immune systems have difficulty figuring out the best attack plan, which makes them feel flulike. When that's the case, the vaccination is even more important; had they been exposed to the real virus, they'd likely become very ill. Every year in the U.S., about 30,000 people die of the flu.

❧ ... And, a duck's quack *does* echo.

Uranus

◆ Discovered in 1781 by amateur astronomer William Herschel, Uranus was the first planet discovered with a telescope! Before his discovery, the only known planets were the six planets one could see with the naked eye.

◆ A gas giant and the seventh planet from the Sun, Uranus is unique in our solar system, as it spins on its side. The scientific jury is still out as to why this is the case.

◆ There are two prevailing theories as to why Uranus spins on its side: either something really gigantic crashed so hard into the planet that it literally knocked it off its axis; or, that the gravitational pull of a really big moon, which was being pulled by an even bigger planet that no longer exists, caused this.

◆ Uranus is dreamy blue-green in color, has 27 known moons, and has a complicated—albeit faint—ring system that includes partial rings and ring arcs. The rings are on their side, too, and encircle Uranus from top to bottom!

◆ Uranus is extremely cold and is often referred to as the "ice giant." The temperature on Uranus is a shockingly chilly -328°F (-200°C).

◆ Uranus is comprised of ice and rock and is 1.7 billion miles (2.7 billion km) from the Sun.

◆ Uranus's atmosphere is primarily comprised of hydrogen and helium, followed by ammonia and methane.

❧ The lower layer of Uranus's clouds is believed to be made of water and the upper layer made of methane. Incidentally, it's the methane that gives Uranus its color.

❧ It has been suggested that Uranus's planetary pressure is so high that it may be covered in, or filled with, enormous diamonds. Now that's the kind of "ice" I'd like to see!

❧ The revolution period of Uranus, or how fast it revolves around the Sun, is about 84 of our Earth years.

❧ What's in a name? Uranus was a god of the sky, the husband of Gaea (Mother Earth) and the father of the Cyclopes, the Titans, Aphrodite, and others.

Vanadium

🐦 A silvery gray, malleable metal with the symbol "V" on the periodic table, vanadium was first discovered in 1801 by Andrés Manuel del Río, who named it *erythronium* (Greek for "red") because when heated, it turned red.

🐦 However, the scientific community disagreed that he had discovered a new element; they believed it was chemically identical to chromium. Andrés Manuel del Río believed them ... although they were mistaken.

🐦 It was "rediscovered" by Nils Gabriel Sefström in 1831, who named it "vanadium" after Vanadís, the Norse goddess of beauty.

V 🐦 Vanadium isn't found as a singular, pure metal; it's found in about 65 different minerals and in fossil fuel deposits.

🐦 South Africa, China, and Russia produce most of the world's vanadium.

🐦 Vanadium pentoxide is used as a catalyst for the commercial production of concentrated sulfuric acid, as well as in the manufacture of ceramics.

🐦 The 1908 Ford Model T car had a chassis made of a vanadium-steel alloy!

Variola Virus
(Smallpox)

🐦 Caused by the variola virus, smallpox is a serious, contagious, and sometimes fatal infectious disease.

🐦 There is evidence that smallpox emerged in human populations around 10,000 BC during the first agricultural settlements in northeastern Africa.

🐦 There is no specific treatment for smallpox disease, and the only prevention is vaccination.

🐦 Evidence of smallpox lesions has been found on the faces of mummies (1570 to 1085 BC), and in the well-preserved mummy of Ramses V, who died as a young man in 1157 BC.

🐦 The way smallpox is portrayed in the movies and on television suggests that people develop symptoms of smallpox right after exposure to someone who is contagious. In reality, there's actually an asymptomatic (symptom-free) incubation period that ranges from seven to 17 days (12 to 14 days is the average time frame).

V

🐦 Not only is it NOT too late to get a smallpox vaccine after being exposed to an infected person, getting vaccinated within three days of exposure will completely prevent or significantly modify smallpox in the vast majority of people. Getting vaccinated as late as four to seven days after exposure likely provides some protection or may modify the severity of disease.

🐦 As a result of the terrorist attacks of 9/11, the U.S. government has enough smallpox vaccine to vaccinate every person in the United States in the case of a smallpox emergency.

🐦 During the French-Indian War (1754 to 1767), the commander of the British forces in North America—Sir Jeffrey Amherst—suggested deliberately infecting the American Indian population with smallpox.

🐦 In 19th-century Europe, about 400,000 people died each year from smallpox.

🐦 There's compelling research that suggests smallpox was responsible for the decimation of the ancient Aztec people.

🐦 The word "variola" comes from the Latin word *varius* or *varus*, meaning "stained" or "mark on the skin." [*See* Smallpox Vaccine History, *page 218.*]

Venus

🐦 Aside from the Sun and Moon, the brightest object in the night sky, and the second closest planet to the Sun, is Venus. Venus is only slightly smaller than Earth, at 7,513 miles (12,092 km) in diameter, only 403 miles (650 km) less than Earth's.

🐦 Although Venus is called Earth's sister planet because its size, chemistry, gravity, and density are almost identical to that of Earth, she's actually our evil twin. The toxic atmosphere and voracious temperatures of Venus would be lethal to any living thing from Earth.

🐦 The hottest planet in our solar system at a stoking 932°F (500°C), the surface of Venus could melt lead!

🐦 Venus is covered with raging active volcanoes, and her thick clouds of sulfuric acid produce the most corrosive acid rain in our entire solar system.

🐦 With an atmosphere of mostly carbon dioxide (96%), plus nitrogen, carbon monoxide, argon, sulfur dioxide, and water vapor, the atmosphere is so heavy (about 90 times heavier than ours) that a person would be crushed instantly. Think of the pressure one would experience 3,000 feet (900 m) down into the ocean, entirely unprotected. Pretty scary.

🐦 Just 67 million miles (108 million km) from the Sun, Venus's revolution period is just under seven and a half months.

🐦 What's in a name? The only planet named after a female, Venus is the goddess of beauty, love, prosperity, and military victory.

V

V-Formation

V

🦢 Besides looking uber-cool, geese (and fighter pilots) fly in that V-shape formation for aerodynamic, communication, and energy-conservation purposes.

🦢 Each bird in the "V" flies slightly above the bird in front of him, resulting in a reduction of wind resistance, which is a great way to both conserve and optimize energy.

🦢 How does that work? Well, each follower bird derives energy from the flow field generated by the next bird ahead in stepped formation.

🦢 The result for the individual birds is lower induced drag, which allows for reduction in the energy required to maintain a given speed.

🦢 The V-formation also helps the flock fly farther before they need to rest as each bird expends less energy than a single bird could achieve flying solo.

🦢 Because the V-formation leader doesn't enjoy the benefit of diminished wind resistance, the birds take turns being the leader then return to the rank and file when they're too tired to lead.

🦢 This benefit of "follow the leader" is enjoyed by fighter pilots via the airflow from their plane's wingtips. That airflow provides energy, and thereby more efficient flight, to another plane flying in an optimum position behind the leader!

In fact, NASA's Ames Research Center found that planes flying in V-formation have a 15% fuel savings.

The V-formation also facilitates the ease with which birds can keep track of each other. No bird left behind!

Flying in the V-formation also assists the flock with communication and coordination within the group. And yes, fighter pilots often use this formation for the same reason!

Vipers

❧ Vipers are venomous snakes, but not poisonous, with two exceptions.

❧ What's the difference between venomous and poisonous? The toxin delivery system. Venom is injected, usually via fangs. Poison is inhaled, eaten, or touched.

❧ What are the two exceptions? Keelback snakes (genus *Rhabdophis*) are poisonous, because they appropriate their toxins from eating poisonous toads, which they secrete via nuchal glands. There's also a garter snake (in the peaceful State of Oregon, of all places) that safely eats poisonous newts, which makes their saliva toxic to some amphibians and other small animals ... and can cause swelling in some people.

❧ Considered by some to be the most advanced snakes due to their sophisticated hollow-fang venom-delivery system, viper fangs are hinged, allowing them to be folded back when not in use.

❧ All the vipers found in North America are in the subfamily of pit vipers (Crotalinae) and have a pair of heat-sensing pits located between each eye and nostril.

V

🐌 Generally speaking, non-venomous snakes have:

- 🐚 Round pupils
- 🐚 No sensing pit
- 🐚 A head slightly wider than their neck
- 🐚 A divided anal plate
- 🐚 A double row of scales on the underside of the tail

🐌 Generally speaking, venomous snakes have:

- 🐚 Elliptical pupils
- 🐚 A sensing pit between eye and nostril
- 🐚 A head much wider than their neck
- 🐚 A single anal plate
- 🐚 Single scales on the underside of the tail

🐌 Snakes are not spineless! Snakes are vertebrates and have between 100 to 400 vertebrae. They also have two long lungs, a long liver, kidneys (one in front of the other), and intestines.

Volcanoes

❧ More than 80% of Earth's surface, above and below sea level, is of volcanic origin.

❧ Three times the height of Mount Everest, the largest volcano in the solar system and the largest mountain in the solar system are one in the same: Olympus Mons on Mars!

❧ Part of the Pacific Ring of Fire, Mount Erebus, located in Antarctica, is the world's southernmost historically active volcano. In Greek mythology, Erebus was a god of darkness, and the son of Chaos.

❧ In descending order, the most common gasses emitted by volcanoes on Earth are water, carbon dioxide, sulfur dioxide, hydrogen chloride, and hydrogen fluoride.

❧ Mauna Loa in Hawaii is believed to be the largest volcano on Earth. It rises 13,000 feet (3,960 m) above sea level and about 29,000 feet (8,840 m) above the seabed.

❧ Estimated to be about 350,000 years old, the oldest volcano on Earth is believed to be Mount Etna, on the island of Sicily, Italy.

❧ Mostly located along the edges of continents, island chains, or beneath the sea forming long mountain ranges, over 50% of Earth's active volcanoes are above sea level, encircling the Pacific Ocean, which forms the Pacific Ring of Fire.

V

The Yellowstone Caldera basin, more commonly referred to as the Yellowstone Supervolcano, is one of the most active volcanoes in the world.

◦❧ Unlike mountains, which are formed by folding and crumpling or by uplift and erosion, volcanoes are built by the accumulation of their own eruptive products, namely lava, volcanic bombs, ashflows, and tephra (which are airborne dust and ash).

◦❧ Ejected in a semisolid, or "plastic," state, volcanic bombs have some rather unusual terms to describe their shapes once they've hit the ground, like "cow dung," "spindle or fusiform," "ribbon," and "bread crust!"

Water

"Water, water, everywhere, nor any drop to drink."
—*Rime of the Ancient Mariner*, by Samuel Taylor Coleridge, 1798

The water cycle on Earth is essentially a closed system, meaning we always have the same amount of water.

Frozen water (ice) takes up about 9% more space than water. A liter of water frozen into ice wouldn't fit back into a liter bottle because it expanded. Ice is also less dense and lighter than water—why ice (and icebergs) float.

In the United States, the rule of thumb is 1 inch (2.5 cm) of rainwater equals about 13 inches (33 cm) of snow. However, the resulting snow total can vary a lot, from 2 inches (5.1 cm) for sleet, 4 inches (10.2 cm) for wet heavy snow, and up to 50 inches (127 cm) for very dry, powdery snow.

Just 1 inch (2.5 cm) of rain falling on one acre (0.004 km²) is equal to 27,154 gallons (102.206 liters) of water and weighs about 113 tons!

Over 70% of Earth's surface is covered by water: 97.5% is salt water and 2.5% fresh. Less than 1% of that fresh, potable water is easily accessible. The majority—about 68.7%—is frozen in ice caps and glaciers; the rest is in the air, soil, or extremely deep aquifers.

The triple point of water is the temperature (273.16°K/0.01°C) and pressure (0.6117 kPa) at which water simultaneously exists as a gas (vapor), liquid (water), and solid (ice). This data is used to calibrate thermometers.

Water Cycle

Evaporation, condensation, and precipitation are the three main elements of the water cycle.

❧ Evaporation is the process by which the liquid water in the oceans and in the soil is transformed into water vapor, an invisible, odorless gas.

❧ Condensation is the process where gaseous water vapor is transformed back into liquid water and forms cloud droplets. As the water vapor rises in the atmosphere, atmospheric temperature decreases with altitude and condensation begins, resulting in the formation of tiny cloud droplets. The tiny cloud droplets begin to collide and coalesce with neighboring cloud droplets, growing in size and weight and eventually forming precipitation.

❧ Precipitation is what "falls" out of the atmosphere as liquid water droplets—rain—or solid water particles—snow and hail.

Weta

- Best described as a nocturnal grasshopper, the weta's genus of *Deinacrida* and *Hemideina* are some of the world's most ancient insect species that exist today.

- Weta outlived dinosaurs and made it through multiple ice ages. Fossil records from the Triassic period reveal that weta have barely changed in 190 million years.

- In New Zealand there are more than 70 endemic species of weta, with some reaching mammoth insect proportions.

- People in the Hauraki Gulf call giant weta "wetapunga" after the god of bad looks. In the South Island, wetas are known as "taipo," which means "demon."

- The heaviest insect in the world is the rat-size giant weta, which grows up to 6 inches (15 cm) long. The record-holding wetapunga (*Deinacrida heteracantha*) weighed an astonishing 2.5 ounces (71 g). Your average worker honeybee weighs a paltry 0.003 oz (0.09 g)!

- Named "jaws," carnivorous wetas with tusks were discovered in 1970.

- And the coolest weta factoid of all? Twentieth-century naturalist Sir Walter Buller wanted some weta specimens for his collection so he submerged one living weta in nearly-boiling water. He tried to drown another by keeping it underwater for four days, but neither one died!

Whales

🐋 There are about 80 species of whales in the world. Estimates range from 79 to 84 distinct species of whales.

🐋 Called "calves" when they're babies, whales usually hang out in groups, feeding and taking care of each other's calves together.

🐋 A group of whales is usually called a pod. However, other terms include school, gam, float, herd, mob, run, troop, or shoal.

🐋 Many whales have no teeth. Instead, they have baleen—comprised of fringe or comblike bone—which they use to filter small crustaceans and other creatures from the water.

🐋 Commercial whaling—which involves catching and usually killing the whales—was banned in 1986 by the International Whaling Commission, but whaling still continues in places such as Canada and Japan. However, some of the whaling falls under the auspices of research—some lethal and some non-lethal.

🐋 Whales were mentioned in the Bible, in Genesis 1:21: "And God created great whales, and every living creature that moveth, which the waters brought forth abundantly, after their kind, and every winged fowl after his kind: and God saw that *it was* good."

Wildfires

- The earliest evidence of a wildfire dates back 450 million years, during the Silurian period.

- Wildfires can zoom along at speeds of upward of 14 mph (23 km/h) and often consume everything like trees, brush, homes, people, animals, and pretty much whatever else is in its way.

- Wildfires have blazed on every continent in the world except Antarctica.

- In January 2003, a bush fire in New South Wales and the Australian Capital Territory ravaged a region almost the size of Texas!

- The primary natural causes of wildfires are lightning strikes, volcanic eruption, sparks from rock falls, and spontaneous combustion.

- The deadliest wildfire in U.S. history occurred in Peshtigo, Wisconsin, on October 8 to 14, 1871. The reason you probably never heard of it is because it happened at the same time as the more-publicized Great Chicago Fire, which occurred during the same stretch of days, from October 8 to 10, 1871.

- Every year in the United States, there are more than 100,000 wildfires that devastate four to five million acres (16,200 to 20,200 km²) of land.

❧ Secondary effects of wildfires, including erosion, landslides, introduction of invasive species, and changes in water quality, are often more disastrous than the fire itself.

❧ Wildfires are a growing natural hazard, posing a threat to life and property, particularly where native ecosystems meet developed areas.

Wines
(Red and White)

"Wine is the healthiest and most health-giving of drinks."
—Louis Pasteur, French microbiologist and chemist (1822 to 1895)

~ The first written accounts of wine date back to third millennium BC in an ancient Sumerian text called the *Epic of Gilgamesh.*

~ Wine is produced by yeast converting sugar to alcohol, CO_2, and heat.

~ When we talk about wine, we're generally talking about wine from *Vitis vinifera* grapes.

~ Wine can be produced from any fruit juice containing fermentable sugars, like apples, peaches, and kiwis.

~ **White wine**

~ White grapes typically go directly to the press after harvesting, which is called the "crush." [*See* Oenology, *page 168.*] The press separates the juice from the skins, seeds, and stems. The resulting juice is moved to a tank, where it settles overnight. The clean juice is racked off of the solids at the bottom of the tank. Yeast is added to clean, racked juice, and the fermentation process begins, which usually takes five to seven days.

- Depending on the type and variety of wine style, secondary fermentation may be induced, called "malolactic fermentation." That process converts malic acid to lactic acid, which is the principle acid in dairy products, and produces a byproduct called diacetyl, which smells like butter! This explains why some chardonnays have a buttery, creamy taste. Then, the wine is aged, blended, filtered, bottled, aged again, and then enjoyed!

Red wine

- Crushed to separate the stems and to break the skins (as the red color comes from the skins), the resulting product is called "must," which is then transferred to a tank and inoculated with yeast for fermentation. This generates carbon dioxide, heat, and alcohol. The carbon dioxide and heat form a "cap," comprised of skins, which rise to the top of the tank.
- The color of red wine comes from the skins, which is achieved by pumping the liquid over the cap, or by punching down the cap into the liquid.
- At the end of fermentation, the resulting red wine is moved to a tank. The skins are removed and pressed to extract more wine to possibly be blended back later.
- Red wines undergo a malolactic fermentation, as it provides some round mouth feel. Then they go to barrel, age, blended, bottled, aged, and then enjoyed.

Wood Storks

🐦 Known as "the flinthead" because of its gray-black, featherless head and neck, the wood stork (*Mycteria americana*) is an endangered species. Back in the 1930s, there were about 20,000 nesting pairs. Today, it's about 8,000 pairs.

🐦 Wood storks fly with their legs and necks outstretched. These impressively large, long-legged wading birds have a glorious wingspan of 5 to 5½ feet (1.5 to 1.7 m) and stand about 45 inches (114 cm) tall.

꙰ When young, these birds have rather dingy gray feathers on their head and a yellowish bill. Once mature, their 6 to 9 inches (15 to 23 cm) bills are black, thick at the base, and slightly decurved; their plumage, sans their head and neck, is snowy white.

꙰ Traveling up to 80 miles (130 km) to feed in shallow waters, wood storks employ a specialized technique for capturing their prey known as "grope-feeding," or "tacto-location." For example, fish need only barely touch the wood stork's partially open bill, and in a matter of just 25 milliseconds the wood stork snaps it bill shut, making it one of the fastest reflexes known in vertebrates!

꙰ The wood stork was once native to places across the southern United States, but the wood stork currently can only be found in Florida, South Georgia, and a small portion of South Carolina.

꙰ Wood storks are birds of freshwater and brackish wetlands, primarily nesting in cypress or mangrove swamps and feeding in freshwater marshes, flooded pastures, and flooded ditches. Thus, one of the reasons for their decline is likely due to the loss of these vital wetland habitats.

World Wide Web

🐦 In 1989, an English physicist named Sir Tim Berners-Lee invented the World Wide Web.

🐦 The letters "HTTP" stand for hypertext transfer protocol.

🐦 In 1945, Vanevar Bush introduced the idea of hypertext.

🐦 The term "hypertext" was actually coined by Ted Nelson.

🐦 The first proposal for a "large hypertext database with typed links" was written by Tim Berners-Lee in 1989, but it generated little interest at the time—outside of the particle physics scientific community, that is!

🐦 The moniker "URL," a website address, stands for uniform resource locator.

🐦 To date, there are over one trillion unique URLs.

🐦 There are over two billion web users across the globe, and that number is growing exponentially every year.

🐦 In 1992, the first photo ever uploaded on the web was an image of the CERN (house band, *Les Horribles Cernettes*). Note: CERN (*Conseil Européen pour la Recherche Nucléaire*) is the European Council for Nuclear Research. They had an in-house all-female musical group of CERN employees, called *Les Horribles Cernettes*. The photo is of the four female members of the band!

Xenobiology

∿ A really cool subfield of astrobiology is xenobiology, which literally means "biology of the stranger."

∿ While astrobiology encompasses the possible existence, nature, and search for extraterrestrial life—or life beyond Earth—and includes elements of biology, astronomy, and geology, xenobiology is the study of possible alternative biochemistries that might be found in extraterrestrial life.

X

🐦 Even though there are not officially any "little green men" to study (yet), there's a great deal of scientific activity going into the search for possible signs of current or former life on Mars, such as robotic missions from NASA and the European Space Agency, in collaboration with the Russians.

🐦 NASA launched the Mars Science Laboratory in 2011. Its Curiosity Lander should arrive on the surface of Mars on August 6, 2012. The mission payload includes instrumentation that will search for possible signs of life, like biosignatures, by studying the gases present in the Martian atmosphere and the composition of minerals.

🐦 Some xenobiologists do their work by studying the unusual biochemistries of microbial organisms found in extreme environments, such as volcanic vents on the ocean floor, to understand what sorts of environments are capable of sustaining life.

🐦 It has long been speculated that although life on Earth uses, almost exclusively, left-handed (L-form) amino acids, life could just have easily evolved using right-handed (D-form) amino acids ... and it might have done so on some other planets in the galaxy! [*See* Amino Acids, *page 1.*]

Xenon

🔖 From the Greek word *xenos*, meaning "stranger," "foreigner," or "guest," xenon occurs in trace amounts in Earth's atmosphere.

🔖 Xenon was discovered in 1898 by Morris Travers and William Ramsey, the same two chemists who discovered the elements neon and krypton. They found xenon in the leftover residue following their evaporating of the components of liquid air!

🔖 Xenon has an atomic number 54 on the periodic table. Today this rare, chemically inert, colorless, tasteless, odorless, heavy, noble gas is obtained commercially by fractional distillation of liquid air. In a nutshell, this involves separating air into oxygen and nitrogen, which yields, in part, liquid oxygen, which contains small quantities of krypton and xenon.

X

🔖 Xenon has been used commercially for decades. The first solid-state laser was pumped by a xenon flash lamp back in 1960. And on a "lighter" note, xenon is also used in stroboscopic lamps—better known as strobe lights!

🔖 In 2006, a study found that using a xenon chloride excimer laser was an effective treatment for a form of psoriasis.

🔖 Xenon can be used as a general anesthetic. In fact, a 2011 study titled "Early Cognitive Function, Recovery and Well-Being after Sevoflurane and Xenon Anesthesia in the Elderly: A Double-Blinded Randomized Controlled Trial" found that emergence from general anesthesia was faster in those from the xenon arm of the study.

<svg> Xenon is currently being studied for its ability to protect brain cells, following various brain injuries such as acute ischemic stroke and traumatic brain injury.

X-Ray

🐦 A physically-painless procedure, getting an X-ray involves electromagnetic radiation to create images of organs and other structures inside the body.

🐦 Shorter in wavelength than UV (ultraviolet) rays, and longer than gamma rays, X-rays have a wavelength ranging from 0.01 to 10 nanometers.

🐦 Contrary to popular belief, X-rays are actually visible to the dark-adapted naked eye, appearing as a faint blue/gray glow.

🐦 In the United States, about 50% of the X-ray radiation exposure we experience comes from outer space and Earth. The remaining 50% comes from medical imaging.

🐦 X-ray images are recorded on a special film called a radiograph. Today, they're also commonly digitized.

🐦 Body parts appear light or dark because of the different rates that your tissues absorb the X-rays. Calcium in bones absorbs X-rays the most, so bones look white on the radiograph, or fluorescent screen; fat and other soft tissues absorb less, and look gray; air absorbs least, so lungs look black.

X

🪶 A computed axial tomography (CAT) scan is a series of X-rays linked to computer technology. Similarly, mammography is a series of breast X-rays, and a barium enema is a series of bowel X-rays with contrast medium—a dye that helps highlight areas that otherwise would be transparent to X-rays.

🪶 In other countries, the X-ray is called "Röntgen radiation," named after physicist Wilhelm Conrad Röntgen, who won the first Nobel Prize in Physics in 1901.

🪶 "X-Ray" is short for "X-Radiation."

🪶 While Wilhelm Conrad Röntgen is generally credited with discovering the X-ray (he was the first to systematically study them), there were many scientists who did groundbreaking work that preceded him, including Ivan Pulyui, Eugen Goldstein, Philip Lenard, Heinrich Hertz, Nikola Tesla, and Thomas Edison.

Xylocopa Violacea

◆ One of the largest and most common carpenter bees in Europe, the *Xylocopa violacea* makes its nest in dead wood.

◆ Sometimes called the "violet carpenter bee" or the Indian bhanvra, the *Xylocopa violacea* bears a resemblance to the European hornet, which can be highly aggressive, territorial, and may sting without warning.

◆ *Xylocopa violacea*, however, are relatively peaceful insects, primarily only attacking when provoked.

◆ The common name, "carpenter bee," comes from the queen bee's behavior after mating: She tunnels her way through dead wood to make a nursery and lay her eggs.

X

◆ Carpenter bees don't eat wood. They either discard it or sometimes may use small bits to build partitions between cells.

◆ Like all bees, *Xylocopa violacea* are trichromatic (just like people). However, we base our color vision on red, green, and blue. Bees base their color vision on ultraviolet, blue, and green light.

◆ And just for the record, bees don't see the color red. Generally speaking, bees are attracted to red flowers because they can see UV markings on the petals.

◆ That being said, *Xylocopa* bees prefer mostly yellow followed by purplish-white, creamy white, and bluish-white flowers.

CARPENTER BEES

❧ The relationships between *Xylocopa* bees and most of the plant species is mutually beneficial for either food or for effecting pollination, which also means they play a key role in human survival, too.

Xylotoles Costatus

🐛 *Xylotoles costatus*, otherwise known as the Pitt Island longhorn beetle, is a flightless, blackish longhorn beetle, sporting a lovely, variable, green-bronze sheen.

🐛 Called "longhorns" for good reason, their antennae are able to fold back against the body. They average 5.9 to 7.8 inches (15 to 20 cm) long.

🐛 As of 1996, *Xylotoles costatus* was designated extinct. However, a population of approximately 200 *Xylotoles costatus* adults has been identified on South East Island in the Chatham Islands group!

🐛 Not a single *Xylotoles costatus* has been seen on Pitt Island since 1907.

🐛 *Xylotoles costatus* is listed as "nationally critical"—the highest threat ranking available in the New Zealand Department of Conservation threat classification system.

🐛 *Xylotoles costatus* also listed as a highest priority threatened species for conservation action in the New Zealand list of threatened invertebrates.

🐛 The Pitt Island longhorn beetle is classified as endangered on the IUCN Red List. Their primary threats appear to be mice, pigs, and people. Mice and pigs eat them, and in terms of people, well, among other things, we're diminishing their natural habitat via land-use encroachment.

X

Yams

~ In spite of what some people in America and Canada think, a sweet potato (*Ipomoea batatas*) is not a yam.

~ Sweet potatoes are roots, members of the morning glory family, and are rich in beta-carotene. Potatoes are tubers, members of the nightshade family, and have no beta-carotene. Yams (genus *Dioscorea*) are tubers, members of the lily family, and have much less beta-carotene than a sweet potato.

~ In Malaysia and Singapore, the starchy tubers of taro or dasheen (*Colocasia esculenta*) may be called yams, but they are not really yams, either.

~ Weighing up to 154 pounds (70 kg), yams are grown in many parts of the world, like Africa, Asia, Latin America, and the Caribbean, and are cooked in many delicious ways.

~ Yams are particularly important as a food source in West Africa, where 95% of the world's yams are harvested.

~ Certain species of yam, particularly a Mexican variety, have high levels of a chemical called diosgenin, which has been used as a precursor for the synthesis of progesterone, a component of oral contraceptive pills.

Yawning

🐦 Pandiculation is the scientific term for the act of stretching and yawning.

🐦 According to the latest research, yawning and stretching may serve to maintain brain thermal homeostasis, meaning yawning and stretching help keep our brain temperature within optimal functioning parameters. These parameters are very limited; temperatures a few degrees above normal may cause, lead to, or result in illness.

🐦 Humans can begin to yawn as early as 11 weeks following conception. So that means a baby can yawn while in the womb.

🐦 Do we also yawn when we're bored? Yes.

🐦 Most mammals yawn, as do some birds and reptiles.

🐦 The average duration of a human yawn is about six seconds.

🐦 Are yawns really contagious? Yes, scientific research indicates that in people, the contagious effect begins in children as young as one or two years old.

🐦 The hypothalamus is a part of the brain that plays an important role in yawning.

🐦 Not surprisingly, our hypothalamus controls and is involved in key metabolic and autonomic nervous system processes, including body temperature, circadian rhythm cycles, fatigue, sleep, hunger, and thirst.

Y

Yellow-Bellied

❧ A beautiful, medium-size woodpecker, the yellow-bellied sapsucker (*Sphyrapicus varius*) is found in both North and Central America, as well as in the Caribbean.

❧ Popular as pets, the yellow-bellied slider (*Trachemys scripta scripta*) is a land and water turtle.

❧ Feeding on small fish only, the yellow-bellied sea snake (*Pelamis platurus*), is about ten times more venomous than the Egyptian cobra (*Naja haje*).

❧ Related to ground squirrels and prairie dogs, the yellow-bellied marmot (*Marmota flaviventris*) is a type of ground squirrel. It is commonly called the "whistle pig" because it chirps and whistles when alarmed by predators like coyotes, foxes, badgers, and eagles.

❧ A small, insect-eating bird, the yellow-bellied flycatcher (*Empidonax flaviventris*) has a distinct song, which ascends from a low to high pitch and sounds like "*chu-wee!*"

❧ The yellow-bellied climbing mouse (*Rhipidomys ochrogaster*) was just recently rediscovered in May 2010 in southeastern Peru, at elevations of 6,000 feet (1,830 m). There were only two known specimens, collected back in 1901.

Y

❧ A type of marsupial found in Australia, the yellow-bellied glider (*Petaurus australis*) is about the size of a rabbit and can glide up to 490 feet (150 m), and jump up about 330 feet (100 m).

❧ Additional "yellow-bellied" creatures include the yellow-bellied sheath-tailed bat, poison frog, magpie, egret, loon, weasel, albatross, kingfisher, swee, and puffing snake.

Yellow Fever

  ✒ Called "yellow" fever because people infected often become jaundiced, this disease is transmitted by infected mosquitoes.

  ✒ Mosquitoes carry the virus from one host to another, primarily between monkeys, from monkeys to humans, and from person to person.

  ✒ A Cuban scientist named Carlos J. Finlay was the first to propose the connection between mosquitoes and yellow fever in 1881. However, he was unsuccessful in conclusively proving his theory to the scientific community.

  ✒ It was Walter Reed—working on the United States Army Yellow Fever Commission—who ultimately identified the mosquito as the carrier of the virus.

  ✒ Reed used human subjects in his research, and more importantly he developed what is believed to be the first informed consent form!

  ✒ Yellow fever is an endemic in tropical areas of Africa and Latin America, which have a combined population of over 900 million people. There are about 200,000 cases of yellow fever, causing 30,000 deaths, each year.

  ✒ Yellow fever is difficult to diagnose, especially during the early stages. It is often confused with severe malaria, dengue hemorrhagic fever, leptospirosis, viral hepatitis (especially the fulminating forms of hepatitis B and D), and other hemorrhagic fevers.

Y

～ The number of yellow fever cases has increased over the past two decades because of declining population immunity to infection, deforestation, urbanization, population movements, and climate change.

～ There is no cure for yellow fever. Treatment is symptomatic, aimed at reducing the symptoms for the comfort of the patient.

～ Vaccination is the most important preventive measure against yellow fever.

～ The vaccine is safe, affordable, and highly effective, and appears to provide protection for 30 to 35 years or more; and probably for life.

～ The vaccine provides effective immunity within one week for 95% of persons vaccinated. Serious side effects from the vaccine are extremely rare.

～ The Word Health Organization (WHO) is the secretariat for the International Coordinating Group for Yellow Fever Vaccine Provision (ICG). The ICG maintains an emergency stockpile of yellow fever vaccines to ensure rapid response to outbreaks in high-risk countries.

Yoga

"Yoga exists in the world because everything is linked."

—T.K.V. Desikachar

❧ A male who practices yoga is called a "yogi," or "yogin." A female who practices yoga is called a "yogini."

❧ Originating in India over 5,000 years ago, yoga is difficult to define. Today, there are so many forms of yoga. That said, at its most basic level, yoga combines a distinctly Eastern philosophy with physical postures, focused breathing, and meditation.

❧ The term "yoga" comes from the Sanskrit word *yui*, meaning, "to unite."

❧ Hatha yoga, commonly practiced in the United States and Europe, emphasizes postures, breathing exercises, and meditation. Examples of Hatha yoga styles include: Ananda, Anusara, Ashtanga, Bikram, Iyengar, Kripalu, Kundalini, and Viniyoga.

❧ Recent studies of people with chronic low-back pain suggest that a carefully adapted set of yoga poses can help reduce pain and improve function, particularly the ability to walk and move.

❧ A 2011 study of 313 adults with chronic or recurring low-back pain suggests that practicing yoga for 12 weeks resulted in better function than usual medical care.

Y

꧁ A small study of 90 people with chronic low-back pain found that participants who practiced Iyengar yoga had significantly less pain and depression and more mobility after six months.

꧁ Numerous studies also suggest that practicing yoga, as well as other forms of regular exercise, may help relieve anxiety and depression, and may also be a useful tool for reducing heart rate and blood pressure.

꧁ *Question:* What did the "Help Wanted" advertisement say for the Yoga Studio searching for a new instructor? *Answer:* Inquire within.

꧁ According to a National Health Interview Survey (NHIS) of Americans, more than 13 million adults and three million children practice yoga.

Yogurt

🐦 From the Turkish word *yoghurmak*, meaning "to thicken," yogurt is rich in protein, calcium, riboflavin, and vitamins B6 and B12. Yogurt is one of the best "feels like you're being bad" foods, but it's really very good for you.

🐦 Pliny the Elder (23 AD to 79 AD) wrote that certain "barbarous nations" knew how "to thicken the milk into a substance with an agreeable acidity."

🐦 The ancient people of Assyria called yogurt *lebeny*, which in Assyrian means "life."

🐦 The combination of yogurt and honey is called "the food of the gods" in ancient Indian records.

🐦 Yogurt was originally marketed as medicine and sold in pharmacies.

🐦 Isaac Carasso of Barcelona first introduced yogurt to Europe.

🐦 In 1919, Carasso founded the company Groupe Danone, which we know as Dannon in the United States. Named for his son, "little Daniel," Daniel inherited the company and expanded the family business to France and the United States.

🐦 The Radlická Mlékárna dairy located in Prague patented the concept of adding fruit jam to yogurt in 1933. The first time fruit was added to commercially produced yogurt was in 1946.

Y

❧ In just 1 cup (236 g) of plain low-fat yogurt, you get about 16% of the daily value of potassium, which, if eaten regularly, can lower blood pressure or hypertension, which, in turn, reduces the formation of blood clots, artery-clogging plaque, and aneurysms.

❧ A really impressive study found that people with high cholesterol who ate a daily serving of yogurt (200 g, or just under 1 cup) containing the live, active culture *L. acidophilus* achieved a 2–3% reduction in their serum cholesterol levels! That's a big deal because every 1% reduction in serum cholesterol concentrations has been shown to confer a 2–3% reduction in estimated risk for coronary heart disease.

❧ A 12-week study found that obese people who included yogurt in their diet lost significantly more fat—a whopping 61% more—and maintained 31% more lean muscle mass than those obese participants who did not include yogurt in their diet.

❧ Perhaps even more significant, those on the yogurt diet lost 81% more abdominal fat. Studies show that body fat location, specifically belly fat, plays a significant role in the development of atherosclerosis, myocardial hypertrophy (an enlarged heart due to working too hard), and coronary heart disease.

❧ About 4.2 billion pounds (2 billion kg) of yogurt is produced in the United States each year.

❧ On average, Americans eat about 12 pounds (5.4 kg) of yogurt per person a year. On average, Swedes consume about 63 pounds (28.5 kg) per person annually. (So, does that mean those of us who regularly eat yogurt are more cultured?)

Yutyrannus Huali
(Fuzzy, Feathered Dinosaurs)

🐦 Scientists from the Institute of Vertebrate Paleontology and Paleoanthropology in the Chinese Academy of Sciences in Beijing discovered a new species of tyrannosauroids, which are fuzzy and feathered!

🐦 An article about the *Yutyrannus huali* was recently published in the very prestigious scientific journal *Nature*.

🐦 The well-preserved remains of three *Yutyrannus huali* skeletons were recently discovered—one full-grown adult and two juveniles.

🐦 The adult remains measured about 30 feet (9 m) from nose to tail and weighed in at an impressive 3,000 pounds (1,360 kg).

🐦 The two younger *Yutyrannus huali* weighed in at "only" 1,300 pounds (589 kg) and 1,100 pounds (498 kg)!

🐦 Beautifully preserved fossilized feathers, ranging in length from 6 to 8 inches (15 to 20 cm), were found on the dinosaurs' neck, hips, feet, arms, and tail!

🐦 Since we know that fossilized birds with full plumage are commonly found with various areas of missing feathers, scientists believe that the *Yutyrannus huali* may have also once been entirely covered with feathers.

Y

🐦 Since dinosaurs are cold-blood, scientists believe that the feathers provided a significant source of insulation.

🐦 Much to the chagrin of naysayers, the discovery of the *Yutyrannus huali* gives additional credence to the theory that theropods, like Giganotosaurus, Spinosaurus, and Tyrannosaurus Rex, are indeed the ancestors of modern birds.

🐦 Theropods share many characteristics with birds. They both have air-filled bones, three-toed feet, and they both brood their eggs—meaning they sit on them to keep them warm as opposed to just laying them and moving on. Both also have a wishbone-shaped collarbone (called a "furcula").

🐦 Wondering why the Chinese scientists called this new species of fuzzy, feathered dinosaurs *Yutyrannus huali*? In Mandarin, the word *Yu* means "feather," and the word *huali* means "beautiful." Great nomenclature!

Zoetropes

🐦 A zoetrope is a primitive form of animated movie created by placing a sequential series of pictures inside a cylinder with narrow slits and viewing through the slits as the cylinder is turned.

🐦 The zoetrope was eventually eclipsed in popularity by the praxinoscope because the image quality was superior. The praxinoscope uses the same sequence of images inside a cylinder but instead of viewing through slits in the side of the drum, one looks down at a centrally located hub of specially placed mirrors.

🐦 Zoetropes do not need to be circular.

Z 🐦 A linear zoetrope can be formed by placing slits in a wall in front of images and moving past the wall at the proper speed. Linear zoetrope can be of any length.

🐦 A mutoscope, which isn't a zoetrope but is similar to one, works when images mounted on a rotating cylinder are flipped in quick succession. These are the old-time, coin-operated, penny arcade "peep show" machines!

🐦 The mechanisms for creating the illusion of motion from still pictures for zoetropes, praxinoscopes, and mutoscopes rely on the phenomenon commonly, but inaccurately, referred to as "Persistence of Vision."

What's actually going on has to do with the psychophysiology of visual perception and the brain's ability to create a perceived reality from the visual input from the optic nerve. In other words, our brains fill in the gaps so what we "see" coalesces with what we know to be possible or true. This explains why we're so easily susceptible to illusions: We like our world to make sense!

Zombie Ants

⟡ "Zombie fire ants" are fire ants that have been infected by a parasitic South American phorid fly.

⟡ The flies lay eggs in the ant's thorax, which then hatch. The larvae then consume the ant from the inside out.

⟡ Eventually the larvae migrate to the ant's head, resulting in peculiar behavior where the ant wanders aimlessly for weeks until its head falls off and the fly hatches to go find more fire ants to zombify.

⟡ Wandering away from the nest probably keeps the infected ants (and parasite) from being attacked by the other ants as a threat.

Z

Zombie Computers

❧ In computer science, a "zombie computer" is an internet-connected computer that has been infected and taken over by a malicious software program like a virus.

❧ This infection causes the zombie computer to respond to the commands of a hacker's remote control network.

❧ This network of zombies (called a "botnet") is used by hackers to send out spam e-mails and/or conduct distributed denial-of-service attacks on target systems.

Z

Zombie Etiquette

  ❧ Zombie Etiquette is a media, arts, and entertainment program on Princeton Community TV, hosted by zombies.

  ❧ Not just for zombies, the program is intended for a broader audience.

  ❧ The programming format utilizes undead hosts to "pick the brains" of filmmakers, authors, musicians, and other guests.

  ❧ The program is filmed on "Studio Z," as well as on location.

Z

Zombies

~&~ The zombie of the African/Haitian Vodoun (voodoo) religion is not the brain-gobbling killing machine of late-night movies, although they are "undead." According to the Vodoun religion, zombies are passive, enslaved creatures.

~&~ Zombies share similarities to creatures from other folkloric traditions—like the Jewish dybbuk and golem, the Icelandic draugr, or the Tibetan ro-lang. These creatures are all raised from the dead (or inanimate material) to serve the sorcerer or magician that raised them.

~&~ Wade Davis, a Harvard ethnobotonist, speculates that "zombie powder," a material reportedly used as part of the Haitian zombie creation ritual, may contain the poison tetrodotoxin, derived from the puffer fish.

Z

~&~ Tetrodotoxin causes paralysis and can induce a comalike state. Davis suggests that this mimics the "return from the dead" theme of the Vodoun zombie legend. Other scientists dispute these claims.

~&~ In May of 2011, the U.S. Centers for Disease Control (CDC) published a blog post called "Preparedness 101: Zombie Apocalypse" as a social media publicity stunt. It was a great success: Not only did it get national news coverage, but the post took their Twitter following from 12,000 to 1.2 million people overnight!

🍂 Although presented in a humorous fashion, the CDC post had a serious message that personal disaster preparedness for any disruption of civil society, whatever the cause, requires careful planning for a period of self-sufficiency until life returns to normal.

🍂 As CDC director, Dr. Ali Khan noted: "If you are generally well equipped to deal with a zombie apocalypse, you will be prepared for a hurricane, pandemic, earthquake, or terrorist attack."

Zoonosis

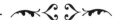

🐦 A zoonosis is any infectious disease that can pass from animals to humans (and sometimes back to the animal).

🐦 A comprehensive literature review by Scottish scientists identified 1,415 species of infectious organisms known to cause human diseases, including 217 viruses and prions, 538 bacteria and rickettsia, 307 fungi, 66 protozoa, and 287 helminths (which are parasitic worms). Out of these, 868 (61%) can be transmitted between humans and animals, and 175 pathogenic species are associated with diseases considered to be "emerging."

🐦 Many zoonoses (plural for zoonosis) are transmitted by a vector, such as a biting insect like a tick, flea, or mosquito.

🐦 The animal population that harbors the infectious agent is called the "reservoir" or "host species."

🐦 Examples of zoonoses include anthrax, bird flu, bubonic plague, cholera, Ebola, Lyme disease, rabies, trichinosis, and tapeworm.

Z

Zooxanthellae

- Zooxanthellae are microscopic algae, or super small plant cells, that live within most types of a tiny marine animal called coral polyps.

- Zooxanthellae enjoy a mutually beneficial relationship with the coral polyps.

- The coral polyps provide the cells with a protected environment, and the vital compounds zooxanthellae need to conduct photosynthesis.

- Zooxanthellae provide the coral polyps with life-sustaining essential nutrients via photosynthesis and help the coral to remove wastes.

Z

- In fact, if the coral polyps go too long without zooxanthellae, the coral can die.

- Here's how it works:
  - Coral polyps produce carbon dioxide and water as byproducts of cellular respiration. [*See* Krebs Cycle, *page 119.*]
  - The zooxanthellae cells use the carbon dioxide and water to carry out photosynthesis, which results in the production of the life-sustaining products such as sugars (glucose), lipids (fats), amino acids, and oxygen.
  - The coral polyp then uses these products to grow and carry out cellular respiration.
  - This extraordinary recycling and exchange of products between the coral polyp cells and the zooxanthellae is the driving force behind the growth and productivity of coral reefs!

Zymurgy

ـ◉ Zymurgy is the science of fermentation processes, such as the brewing of beer.

ـ◉ The oldest definitive evidence of beer-making includes a chemical residue on pottery jars dating to 3,500 BC. There is also evidence that the process existed many years ago in a 3,900-year-old Sumerian poem containing what may be the earliest beer-making recipe!

ـ◉ However, based on the evidence of barley domestication and cultivation found in Neolithic archeological sites across the Fertile Crescent, beer-making likely dates to 8,500 BC. It *could* date back even earlier, based on evidence of wild barley collection at a site near the Sea of Galilee from 17,000 BC.

ـ◉ From 1516 until 1987, a Bavarian law called the Reinheitsgebot restricted the ingredients in beer to water, barley (a grain), and hops (the female flower seed cones from the perennial vine *Humulus lupulus*). The law was later adopted across Germany.

ـ◉ Yeast was added to the "acceptable ingredients" list after Louis Pasteur discovered its vital role in beer-making in 1857. Wild yeast was always part of the fermentation process ... they just didn't know it before Pasteur.

ـ◉ Most brewing is done with the budding yeast *Saccharomyces cerevisiae* ("cerevisiae" means beer in Latin). However, certain traditional African millet-based brews employ the fission yeast *Schizosaccharomyces pombe* ("pombe" is the Swahili word for beer).

❧ Hops also play an important role in the process of beer-making. Not only do they add flavor, but they also act as a preservative. In earlier days, this increased its shelf life and made beer feasible to transport long distances for trade. Today beer is pasteurized. Hops certainly still has preservative capabilities, but its role today is really just about imparting a pleasant bitterness to balance the maltiness.

❧ In the late 1700s, the invention of the thermometer and the hydrometer (for measuring the relative density of liquids) vastly improved the scientific basis of beer-making by allowing precise measurements during the beer-making process, which led to more consistent flavor and quality.

Bibliography

Alberts, B. *Molecular Biology of the Cell* (4th ed.). New York: Garland Science. 2002.

American Cancer Society; Centers for Disease Control (CDC) and Prevention.

Andersen, S. B., Ferrari, M., Evans, H. C., Elliot, S. L., Boomsma, J. J., and Hughes, D. P. "Disease Dynamics in a Specialized Parasite of Ant Societies," *PLoS One*, 7(5) (2012): e36352. doi: 10.1371/journal.pone.0036352.

Bes-Rastrollo, M., Sabate, J., Gomez-Gracia, E., Alonso, A., Martinez, J. A., and Martinez-Gonzalez, M. A. "Nut Consumption and Weight Gain in a Mediterranean Cohort: The SUN Study," *Obesity (Silver Spring)*, 15(1) (2007): 107–116.

Birks, J., and Grimley Evans, J. "Ginkgo Biloba for Cognitive Impairment and Dementia," *Cochrane Database Syst Rev*(1) (2009): CD003120. doi: 10.1002/14651858.CD003120.pub3.

Boguski, M. S. "A Molecular Biologist Visits Jurassic Park," *BioTechniques* 12(5):668–669)(1992).

Bradish, C. M., Perkins-Veazie, P., Fernandez, G. E., Xie, G., and Jia, W. "Comparison of Flavonoid Composition of Red Raspberries (*Rubus idaeus L.*) Grown in the Southern United States," *J Agric Food Chem.* (2011): doi: 10.1021/jf203474e.

Brinkley, T. E., Lovato, J. F., Arnold, A. M., Furberg, C. D., Kuller, L. H., Burke, G. L., Williamson, J. D. "Effect of Ginkgo Biloba on Blood Pressure and Incidence of Hypertension in Elderly Men and Women," *Am J Hypertens*, 23(5) (2010): 528–533. doi: 10.1038/ajh.2010.14.

Bull, J. J., Jessop, T. S., and Whiteley, M. "Deathly Drool: Evolutionary and Ecological Basis of Septic Bacteria in Komodo Dragon Mouths," *PLoS One*, 5(6) (2010): e11097. doi: 10.1371/journal.pone.0011097.

CDC, Office of Public Health Preparedness and Response: Zombie Preparedness 101. *Centers for Disease Control and Prevention, Atlanta, Georgia.* 2012.

Cooke, M., Moyle, W., Shum, D., Harrison, S., and Murfield, J. "A Randomized Controlled Trial Exploring the Effect of Music on Quality of Life and Depression in Older People with Dementia," *J Health Psychol*, 15(5) (2010): 765-776. doi: 10.1177/1359105310368188.

Cotton, F. A. *Advanced Inorganic Chemistry* (6th ed.). New York. Wiley. 1999.

Dahm, R. "Discovering DNA: Friedrich Miescher and the Early Years of Nucleic Acid Research," *Hum Genet, 122*(6) (2008): 565-581. doi: 10.1007/s00439-007-0433-0.

Davidson, A. D., Boyer, A. G., Kim, H., Pompa-Mansilla, S., Hamilton, M. J., Costa, D. P., ... Brown, J. H. "Drivers and Hotspots of Extinction Risk in Marine Mammals," *Proc Natl Acad Sci U S A, 109*(9) (2012), 3395-3400. doi: 10.1073/pnas.1121469109.

Di Pierro, F., Rapacioli, G., Ferrara, T., and Togni, S. "Use of a Standardized Extract from Echinacea Angustifolia (Polinacea) for the Prevention of Respiratory Tract Infections," *Altern Med Rev, 17*(1) (2012); 36–41.

Duthie, C., Gibbs, G., and Burns, K. C. "Seed Dispersal by Weta," *Science, 311*(5767), 1575 (2006); doi: 311/5767/1575 [pii] 10.1126/science.1123544.

Edwards, K., Kwaw, I., Matud, J., and Kurtz, I. "Effect of Pistachio Nuts on Serum Lipid Levels in Patients with Moderate Hypercholesterolemia," *Journal of the American College of Nutrition, 18*(3) (1999): 229–232.

Hu, F. B., and Willett, W. C. "Optimal Diets for Prevention of Coronary Heart Disease," *JAMA, 288*(20) (2002): 2569–2578.

Keen, C. L., Holt, R. R., Oteiza, P. I., Fraga, C. G., and Schmitz, H. H. "Cocoa Antioxidants and Cardiovascular Health," *Am J Clin Nutr, 81*(1 Suppl) (2005); 298S–303S.

Kieser, J. A., Tkatchenko, T., Dean, M. C., Jones, M. E., Duncan, W., and Nelson, N. J. "Microstructure of Dental Hard Tissues and Bone in the Tuatara Dentary, Sphenodon Punctatus (Diapsida: Lepidosauria: Rhynchocephalia)," *Front Oral Biol, 13* (2008); 80–85. doi: 000242396 [pii] 10.1159/000242396.

Kinsel, J. F., and Straus, S. E. Complementary and Alternative Therapeutics: Rigorous Research is Needed to Support Claims," *Annu Rev Pharmacol Toxicol, 43* (2003): 463–484. doi: 10.1146/annurev.pharmtox. 43.100901.135757.

Kontogianni, M. D., Panagiotakos, D. B., Chrysohoou, C., Pitsavos, C., Zampelas, A., and Stefanadis, C. "The Impact of Olive Oil Consumption Pattern on the Risk of Acute Coronary Syndromes: The CARDIO2000 Case-control Study," *Clin Cardiol, 30*(3) (2007), 125–129.

Lee, K. C., Chao, Y. H., Yiin, J. J., Chiang, P. Y., and Chao, Y. F. "Effectiveness of Different Music-Playing Devices for Reducing Preoperative Anxiety: A Clinical Control Study," *Int J Nurs Stud, 48*(10) (2011): 1180–1187. doi: 10.1016/j.ijnurstu.2011.04.001.

Lee, T. M., and Jetz, W. "Unravelling the Structure of Species Extinction Risk for Predictive Conservation Science," *Proc Biol Sci, 278*(1710) (2011), 1329–1338. doi: rspb.2010.1877 [pii] 10.1098/rspb.2010.1877.

Lehninger, A. L., Nelson, D. L., and Cox, M. M. *Lehninger Principles of Biochemistry* (5th ed.). New York: W.H. Freeman. 2008.

Leonard, W. *Color Yourself Smart: Human Anatomy.* San Diego: Thunder Bay Press. 2011.

Leonard, W., Kastan, K., Banfield, S., and WomenHeart. National Coalition for Women with Heart Disease. *WomenHeart's All Heart Family Cookbook: Featuring the 40 Foods Proven to Promote Heart Health.* Emmaus, Pa.: Rodale. 2008.

Letts, L., Minezes, J., Edwards, M., Berenyi, J., Moros, K., O'Neill, C., and O'Toole, C. "Effectiveness of Interventions Designed to Modify and Maintain Perceptual Abilities in People with Alzheimer's Disease and Related Dementias," *Am J Occup Ther, 65*(5) (2011), 505–513.

Maguire, E. A., Gadian, D. G., Johnsrude, I. S., Good, C. D., Ashburner, J., Frackowiak, R. S., and Frith, C. D. "Navigation-related Structural Change in the Hippocampi of Taxi Drivers," *Proc Natl Acad Sci U S A, 97*(8) (2000): 4398–4403. doi: 10.1073/pnas.070039597.

McCusker, C., and Gardiner, D. M. "The Axolotl Model for Regeneration and Aging Research: A Mini-Review," *Gerontology* (2011): doi: 000323761 [pii] 10.1159/000323761.

Meijer, H. J., Gill, A., de Louw, P. G., Van Den Hoek Ostende, L. W., Hume, J. P., and Rijsdijk, K. F. "Dodo Remains from an in situ Context from Mare aux Songes, Mauritius," *Naturwissenschaften, 99*(3) (2012), 177–184. doi: 10.1007/s00114-012-0882-8.

Monaghan, J. R., Walker, J. A., Page, R. B., Putta, S., Beachy, C. K., and Voss, S. R. "Early Gene Expression During Natural Spinal Cord Regeneration in the Salamander Ambystoma Mexicanum," *J Neurochem, 101*(1) (2007), 27–40. doi: 10.1111/j.1471–4159.2006.04344.x.

Nahas, R., and Balla, A. "Complementary and Alternative Medicine for Prevention and Treatment of the Common Cold," *Can Fam Physician, 57*(1) (2011), 31–36.

NASA,"One Small Step: Official NASA Transcript, Audio and Video Recording," *Apollo 11 Lunar Surface Journal* (1995/2009/2011).

NASA. "NASA Jet Propulsion Laboratory," *National Aeronautics and Space Administration* (2012).

NCBI. "Problem Set: Jurassic Park DNA Sequence: GenBank, RefSeq, and Entrez, Structures, BLAST. Genomes," *National Center for Biotechnology Information Field Guide* (2003).

Necker, R., "Head-bobbing of Walking Birds. *J Comp Physiol A Neuroethol Sens Neural Behav Physiol, 193*(12) (2007), 1177-1183. doi: 10.1007/s00359-007-0281-3.

Neri, S., Signorelli, S. S., Torrisi, B., Pulvirenti, D., Mauceri, B., Abate, G., ... Leotta, C., "Effects of Antioxidant Supplementation on Postprandial Oxidative Stress and Endothelial Dysfunction: A Single-blind, 15-day Clinical Trial in Patients with Untreated Type 2 Diabetes, Subjects with Impaired Glucose Tolerance, and Healthy Controls," *Clin Ther, 27*(11) (2005), 1764–1773.

NOAA. Winter Weather Basics: How do winter storms form. *The National Oceanic and Atmospheric Administration's National Severe Storm Laboratory* (2006).

Pieterse, Z., Jerling, J. C., Oosthuizen, W., Kruger, H. S., Hanekom, S. M., Smuts, C. M., and Schutte, A. E. "Substitution of High Monounsaturated Fatty Acid Avocado for Mixed Dietary Fats During an Energy-restricted Diet: Effects on Weight Loss, Serum Lipids, Fibrinogen, and Vascular Function. *Nutrition (Burbank, Los Angeles County, Calif, 21*(1) (2005): 67–75.

Sajeevan, G. "Latitude and Longitude—A Misunderstanding," *Current Science, Vol.94 No.5.*(2008).

Shoup-Knox, M. L., Gallup, A. C., Gallup, G. G., and McNay, E. C. "Yawning and Stretching Predict Brain Temperature Changes in Rats: Support for the Thermoregulatory Hypothesis," *Front Evol Neurosci, 2*, 108 (2010): doi: 10.3389/fnevo.2010.00108.

Snijder, M. B., van der Heijden, A. A., van Dam, R. M., Stehouwer, C. D., Hiddink, G. J., Nijpels, G., ... Dekker, J. M. "Is Higher Dairy Consumption Associated with Lower Body Weight and Fewer Metabolic Disturbances?" The Hoorn Study. *Am J Clin Nutr, 85*(4) (2007): 989–995.

Szeto, Y. T., Tomlinson, B., and Benzie, I. F. "Total Antioxidant and Ascorbic Acid Content of Fresh Fruits and Vegetables: Implications for Dietary Planning and Food Preservation," *Br J Nutr, 87*(1) (2002): 55–59.

Temple, S. A. "Plant-animal Mutalism: Coevolution with Dodo Leads to Near Extinction of Plant," *Science, 197*(4306) (1977): 885–886. doi: 10.1126/science.197.4306.885.

Tjonneland, A., Gronbaek, M., Stripp, C., and Overvad, K. "Wine Intake and Diet in a Random Sample of 48763 Danish Men and Women". *Am J Clin Nutr, 69*(1) (1999), 49–54.

United States Department of Agriculture. *Dairy Products 2011 Summary (p. v.),* Washington, D.C.: U.S. Dept. of Agriculture, National Agricultural Statistics Service. 2011.

University of Michigan. The Water Resources of Earth, *Human Appropriation of the World's Fresh Water Supply; Lecture, University of Michigan,* 1–11. 2006.

USGS, The Water Cycle: Water Storage in Ice and Snow. *United States Geological Survey, United States Department of the Interior.* 2011.

Valtin, H. "Drink at Least Eight Glasses of Water a Day." Really? Is there scientific evidence for "8 x 8"? *Am J Physiol Regul Integr Comp Physiol, 283*(5) (2002): R993–1004.

Wang, Z., Huang, Y., Zou, J., Cao, K., Xu, Y., and Wu, J. M. Effects of red wine and wine polyphenol resveratrol on platelet aggregation in vivo and in vitro. *Int J Mol Med, 9*(1) (2002): 7779.

Woollett, K., & Maguire, E. A. "The Effect of Navigational Expertise on Wayfinding in New Environments". *J Environ Psychol, 30*(4-2) (2010): 565–573. doi: 10.1016/j.jenvp.2010.03.003.